COMMAND AND CONTROL: THE SOCIOTECHNICAL PERSPECTIVE

Human Factors in Defence

Series Editors:

Dr Don Harris, Cranfield University, UK
Professor Neville Stanton, Brunel University, UK
Professor Eduardo Salas, University of Central Florida, USA

Human factors is key to enabling today's armed forces to implement their vision to 'produce battle-winning people and equipment that are fit for the challenge of today, ready for the tasks of tomorrow and capable of building for the future' (source: UK MoD). Modern armed forces fulfil a wider variety of roles than ever before. In addition to defending sovereign territory and prosecuting armed conflicts, military personnel are engaged in homeland defence and in undertaking peacekeeping operations and delivering humanitarian aid right across the world. This requires top class personnel, trained to the highest standards in the use of first class equipment. The military has long recognised that good human factors is essential if these aims are to be achieved.

The defence sector is far and away the largest employer of human factors personnel across the globe and is the largest funder of basic and applied research. Much of this research is applicable to a wide audience, not just the military; this series aims to give readers access to some of this high quality work.

Ashgate's *Human Factors in Defence* series comprises of specially commissioned books from internationally recognised experts in the field. They provide in-depth, authoritative accounts of key human factors issues being addressed by the defence industry across the world.

Command and Control: The Sociotechnical Perspective

GUY H. WALKER,
Heriot-Watt University

NEVILLE A. STANTON,
University of Southampton

PAUL M. SALMON,
Monash University, Australia

&

DANIEL P. JENKINS,
Sociotechnic Solutions, UK

CRC Press
Taylor & Francis Group
Boca Raton London New York

CRC Press is an imprint of the
Taylor & Francis Group, an **informa** business

CRC Press
Taylor & Francis Group
6000 Broken Sound Parkway NW, Suite 300
Boca Raton, FL 33487-2742

First issued in paperback 2017

© 2009 by Guy H. Walker, Neville A. Stanton, Paul M. Salmon and Daniel P. Jenkins
CRC Press is an imprint of Taylor & Francis Group, an Informa business

No claim to original U.S. Government works

Version Date: 20160226

ISBN 13: 978-1-138-07686-0 (pbk)
ISBN 13: 978-0-7546-7265-4 (hbk)

Visit the Taylor & Francis Web site at
http://www.taylorandfrancis.com

and the CRC Press Web site at
http://www.crcpress.com

Contents

List of Figures

List of Tables

Acknowledgements

The Human Factors Integration Defence Technology Centre (HFI DTC) is a consortium of industry and academia working in cooperation on a series of defence-related projects. The consortium is led by Aerosystems International and comprises the University of Southampton, the University of Birmingham, Cranfield University, Lockheed Martin, MBDA and SEA Ltd. The consortium was recently awarded The Ergonomics Society President's Medal for work that has made a significant contribution to original research, the development of methodology and application of knowledge within the field of ergonomics.

Aerosystems International	University of Southampton	Birmingham University	Cranfield University
Dr Karen Lane	Professor Neville Stanton	Prof Chris Baber	Dr Don Harris
Linda Wells	Dr Guy Walker	Professor Bob Stone	Andy Farmilo
Kevin Bessell	Dr Daniel Jenkins	Dr Huw Gibson	Geoff Hone
Nicola Gibb	Dr Paul Salmon	Dr Robert Houghton	Jacob Mulenga
Robin Morrison	Laura Rafferty	Richard McMaster	Ian Whitworth
Dr Carol Deighton			John Huddlestone
			Antoinette Caird-Daley

Lockheed Martin UK	MBDA Missile Systems	Systems Engineering and Assessment (SEA) Ltd
Mick Fuchs	Steve Harmer	Dr Georgina Fletcher
Lucy Mitchell	Dr Carol Mason	Dr Anne Bruseberg
Rebecca Stewart	Chris Vance	Dr Iya Solodilova-Whiteley
	David Leahy	Ben Dawson

We are grateful to DSTL who have managed the work of the consortium, in particular to Geoff Barrett, Bruce Callander, Jen Clemitson, Colin Corbridge, Roland Edwards, Alan Ellis, Jim Squire, Alison Rogers and Debbie Webb. We are also grateful to Dr John Ardis for review and comment on the work that comprises Chapters 4 and 6, and to Kevin Bessell for assistance in Chapter 5.

This work from the Human Factors Integration Defence Technology Centre was part-funded by the Human Sciences Domain of the UK Ministry of Defence Scientific Research Programme. Further information on the work and people that comprise the HFI DTC can be found on www.hfidtc.com.

About the Authors

Dr Guy H. Walker

School of the Built Environment, Heriot-Watt University, Edinburgh, [UK] EH14 4AS

G.H.Walker@hw.ac.uk

Guy Walker has a BSc Honours degree in Psychology from the University of Southampton and a PhD in Human Factors from Brunel University. His research interests are wide ranging, spanning driver behaviour and the role of feedback in vehicles, railway safety and the issue of signals passed at danger, and the application of sociotechnical systems theory to the design and evaluation of military command and control systems. Guy is the author/co-author of over forty peer reviewed journal articles and several books. This volume was produced during his time as Senior Research Fellow within the HFI DTC. Along with his colleagues in the research consortium, Guy was awarded the Ergonomics Society's President's Medal for the practical application of Ergonomics theory. Guy currently resides in the School of the Built Environment at Heriot-Watt University in Edinburgh, Scotland, working at the cross-disciplinary interface between engineering and people.

Professor Neville A. Stanton

HFI DTC, School of Civil Engineering and the Environment, University of Southampton, Southampton, [UK] SO17 1BJ.

n.stanton@soton.ac.uk

Professor Stanton holds a Chair in Human Factors and has published over 140 international peer-reviewed journal papers and 14 books on Human Factors and Ergonomics. In 1998, he was awarded the Institution of Electrical Engineers Divisional Premium Award for a co-authored paper on Engineering Psychology and System Safety. The Ergonomics Society awarded him the President's medal in 2008 and the Otto Edholm Medal in 2001 for his contribution to basic and applied ergonomics research. The Royal Aeronautical Society awarded him the Hodgson Medal and Bronze Award with colleagues for their work on flight-deck safety. Professor Stanton is an editor of *Ergonomics* and on the editorial board of *Theoretical Issues in Ergonomics Science* and the *International Journal of Human Computer Interaction*. Professor Stanton is a Fellow and Chartered Occupational Psychologist registered with The British Psychological Society, and a Fellow of The Ergonomics Society. He has a BSc in Occupational Psychology from Hull University, an MPhil in Applied Psychology from Aston University, and a PhD in Human Factors, also from Aston.

Paul M. Salmon

Human Factors Group, Monash University Accident Research Centre, Monash University, Victoria, Australia [3800]
Paul.Salmon@muarc.monash.edu.au
Paul Salmon is a senior research fellow within the Human Factors Group at the Monash University Accident Research Centre and holds a BSc in Sports Science, an MSc in Applied Ergonomics, and a PhD in Human Factors. Paul has eight years experience of applied Human Factors research in a number of domains, including the military, aviation, and rail and road transport, and has worked on a variety of research projects in these areas. This has led to him gaining expertise in a broad range of areas, including situation awareness, human error, and the application of Human Factors methods, including human error identification, situation awareness measurement, teamwork assessment, task analysis, and cognitive task analysis methods. Paul has authored and co-authored four books and numerous peer-reviewed journal articles, conference articles, and book chapters, and was recently awarded the 2007 Royal Aeronautical Society Hodgson Prize for a co-authored paper in the society's *Aeronautical Journal* and, along with his colleagues from the Human Factors Integration Defence Technology Centre (HFI DTC) consortium, was awarded the Ergonomics Society's President's Medal in 2008.

Dr Daniel P. Jenkins

Sociotechnic Solutions, St Albans, Herts, [UK]AL1 2LW
info@sociotechnicsolutions
Dan Jenkins graduated in 2004 from Brunel University with an M.Eng (Hons) in Mechanical Engineering and Design, receiving the 'University Prize' for the highest academic achievement in the school. As a sponsored student, Dan finished university with over two years' experience as a design engineer in the automotive industry. Upon graduation, Dan went to work in Japan for a major car manufacturer, facilitating the necessary design changes to launch a new model in Europe. In 2005, Dan returned to Brunel University taking up the full-time role of research fellow in the Ergonomics Research Group, working primarily on the HFI DTC project. Dan studied part time on his PhD in Human Factors and interaction design – graduating in 2008 – receiving the 'Hamilton Prize' for the Best Viva in the School of Engineering and Design. Both academically and within industry, Dan has always had a strong focus on Human Factors, system optimization, and design for inclusion. Dan has authored and co-authored numerous journal papers, conference articles, book chapters, and books. Dan and his colleagues on the HFI DTC project were awarded the Ergonomics Society's President's Medal in 2008. Dan now works as a consultant and his company is called Sociotechnic Solutions.

Chapter 1

Introduction

'The [...] military could use battlefield sensors to swiftly identify targets and bomb them. Tens of thousands of warfighters would act as a single, self-aware, coordinated organism. Better communications would let troops act swiftly and with accurate intelligence, skirting creaky hierarchies. It'd be "a revolution in military affairs unlike any seen since the Napoleonic Age." And it wouldn't take hundreds of thousands of troops to get a job done—that kind of "massing of forces" would be replaced by information management. [...] Computer networks and the efficient flow of information would turn [the] chain saw of a war machine into a scalpel.' (Shachtman, 2007; Cebrowski and Gartska, 1998)

Network Enabled Capability

Network Enabled Capability (NEC) is a type of command and control. Command and control is the generic label for the management infrastructure behind any large, complex, dynamic resource system (Harris and White, 1987). Like all good management infrastructures, military command and control is contingent (e.g., Mintzberg, 1979) upon the problem it is tasked with dealing and because that problem, at some fundamental level, has remained relatively stable over a long period of time the term command and control has become a synonym. It has come to mean exactly, or very nearly the same thing as traditional, hierarchical, bureaucratic, centralised, 'classic' command and control. This is the type of command and control conjured up by images of mass battle scenes in the film Alexander, the type afflicted by the Kafkaesque pathologies inherent in all bureaucratising organisations and parodied by everyone from Black Adder to Dilbert, and the type of command and control that management books try to encourage their readers to 'break free from' (e.g. Seddon, 2003). The arrival of NEC shows that not all management infrastructures need to be the same. It shows that while traditional command and control is highly adept in some situations it is less effective in others.

There is no one definition of NEC. In fact even the label NEC is just one in a number of 'net-enabled' acronyms running from Network Centric Warfare, Network Centric Operations (both favoured in North America) to Network Centric Defence (favoured in some North European countries). Regardless of acronym, the techno-organisational vision of NEC (the preferred label in the UK) refers to:

'...self-synchronizing forces that can work together to adapt to a changing environment, and to develop a shared view of how best to employ force and effect to defeat the enemy. This vision removes traditional command hierarchies and empowers individual units to interpret the broad command intent and evolve a flexible execution strategy with their peers.' (Ferbrache, 2005, p. 104)

The logic of this organismic approach to command and control is widely held to be based on four tenets: that a) a robustly networked force improves information sharing, that b) information sharing and collaboration enhance the quality of information and shared situational awareness, that c) shared situational awareness enables self-synchronisation, and that d) these dramatically increase mission effectiveness (CCRP, 2009). These assumptions have been appropriated by the military domain and have gathered momentum, so it is interesting to note that they are not based on much more than a decade's worth of direct military experience. They are based on Wal Mart:

> 'Here was a sprawling, bureaucratic monster of an organisation—sound familiar?—that still managed to automatically order a new light bulb every time it sold one. Warehouses were networked, but so were individual cash registers. So were the guys who sold Wal-Mart the bulbs. If that company could wire everyone together and become more efficient, then [military] forces could, too. "Nations make war the same way they make wealth".' (Shachtman, 2007; Cebrowski and Gartska, 1998)

This characterisation is perhaps a little crude, and perhaps not entirely fair, but it nevertheless helps us to make a powerful point that goes right to the heart of this book. Or rather, it enables Cebrowski and Gartska (1998) in their pioneering paper on the origins and future of NEC to make it for us. They say:

> 'We may be special people in the armed forces, but we are not a special case. It would be false pride that would keep us from learning from others. The future is bright and compelling, but we must still choose the path to it. Change is inevitable. We can choose to lead it, or be victims of it. As B. H. Liddell Hart said, "The only thing harder than getting a new idea into the military mind is getting an old one out".' (Cebrowski and Gartska, 1998, p. 8)

Command and control is an organisation just like any other, conceptually at least. What has happened to make NEC the contingent response? What is it about the current state of affairs that makes NEC's assumptions and tenets make sense? Wal Mart may be the proximal inspiration for at least some of the early thinkers on the topic of NEC but it is probably closer to the truth to say that it derives from a much wider paradigm shift, of which Wal Mart's networked, vertically integrated operations are merely a part. Cebrowski and Gartska looked over what they saw as the artificial and contentious divide between the military arena and the rest of the world and saw that:

> **The underlying economics have changed**. 'The new dynamics of competition are based on increasing returns on investment, competition within and between

ecosystems, and competition based on time. Information technology (IT) is central to each of these.' Economies are 'characterised by extraordinary growth and wealth generation, increasing returns on investment, the absence of market share equilibrium, and the emergence of mechanisms for product lock-in. [...] Locking-out competition and locking-in success can occur quickly, even overnight. We seek an analogous effect in warfare.' (p. 1–2)

The underlying technologies have changed. 'Information technology is undergoing a fundamental shift from platform-centric computing to network-centric computing. Platform-centric computing emerged with the widespread proliferation of personal computers in business and in the home. [...] These technologies, combined with high-volume, high-speed data access (enabled by the low-cost laser) and technologies for high-speed data networking (hubs and routers) have led to the emergence of network-centric computing. Information "content" now can be created, distributed, and easily exploited across the extremely heterogeneous global computing environment.' (p. 2–3)

The business environment has changed. 'First, many firms have shifted their focus to the much larger, adaptive, learning ecosystems in which they operate. Not all actors in an ecosystem are enemies (competitors); some can have symbiotic relationships with each other. For such closely coupled relationships, the sharing of information can lead to superior results. Second, time has increased in importance. Agile firms use superior awareness to gain a competitive advantage and compress timelines linking suppliers and customers. [...] Dominant competitors across a broad range of areas have made the shift to network-centric operations—and have translated information superiority into significant competitive advantage.' (p. 3–4)

Where Did the Humans Go?

The vision just described is that of Alvin Toffler's 'future shock' (1981) and 'third wave' (1980). While it may have been appealing and relevent there were problems with the assumptions being made about information sharing, shared situational awareness, self-synchronisation and mission effectiveness. One could look over the military/civilian divide and see these fundamental shifts not as new form of business ecology at all, but as a new form of ultra-Taylorism, a situation in which net centric technology was being pressed into the service of even greater bureaucratisation. In a lot of cases what the network seemed to be enabling were the latest Japanese business practices such as lean manufacturing, total quality management, just-in-time stock control, and all the other tools and techniques which in a dynamic and changeable commercial world enable companies, like Wal Mart, to keep at least some aspects of their complex operations in a tightly coupled and predictable state. Naturally, whilst this generally held for the benefactors of the techniques like Toyota and Wal Mart, the same could not always be said for the people at work on the production line of product or service delivery, or indeed the users of roads now overloaded with 'just in time' delivery traffic. The 'technical' aspects had clearly been optimised but at the expense of the 'social'. Whilst from some angles these commercial systems looked like an ecosystem of self-organising flexibility, the alternative perspective was that the network had actually enabled a far closer coupled system, the ultimate high speed hierarchy. The original thinking behind NEC bears this hallmark.

It started off with 'short, decisive battles against another regular army' in mind (Shachtman, 2007, p. 5), 'the Soviets, the Chinese, Saddam's Republican Guard, whoever – as long as they had tanks to destroy, territory to seize, and leaders to kill.' (p. 5). When Cebrowski and Gartska discussed Wal Mart in the same breath as warfare what they were really driving at was: 'a single, network-enabled process: killing' (Shachtman, 2007, p. 8). It is no coincidence that the phrase 'shock and awe' (e.g., Ullman and Wade, 1996) made its way into the military lexicon at roughly the same time. But then something started to happen. After the technically optimized military successes had occurred and the mission accomplished banners had been taken down from the bridge of the USS Abraham Lincoln, a new type of organisational pathology emerged, one for which the military context seemed particularly susceptible:

> 'He clicks again, and the middle screen switches to a 3-D map of an Iraqi town from a driver's point of view. "Now let's plan the route. You've got a mosque here. An IED happened over there two weeks ago. Here's the one that happened yesterday. Hey, that's too close. Let's change my route. Change the whole damn thing." He guides me through capability after capability of the command post— all kinds of charts, overlays, and animations. "But wait—there's more," he says. "You wanna see where all the Internet cafés are in Baghdad?" […] It's hard not to get caught up in [the] enthusiasm. But back in the US, John Nagl, one of the authors of the Army's new counterinsurgency manual, isn't impressed. […] he's more interested in what the screens don't show. Historical sigacts don't actually tell you where the next one's going to be. Or who's going to do it. Or who's joining them. Or why. "The police captain playing both sides, the sheikh skimming money from a construction project, […] what color [icon] are they?".'
> (Shachtman, 2007, p. 5)

The paradox in all this is that if, as Cebrowski and Gartska acknowledge, 'military operations are enormously complex, and complexity theory tells us that such enterprises organise best from the bottom-up' (1998, p. 4–5) then what happened to the most important low-level component of them all: the human? The focus on Toffler's third wave of fundamental changes in economics, technology and business has tended to put the focus on the 'network' rather than what it 'enables'. Which is for the most agile, self-synchronous component of all in NEC, the people, to use its capability to perform the one task they excel at: coping with complexity.

For the human factors practitioner encountering the extant literature on NEC for the first time the overriding feeling is undoubtedly one of opportunity. Here we have a huge practical domain of interdisciplinary science where, for once, the human is widely acknowledged to be key. Within this expanding literature are concepts and ideas which have tremendous potential, not only for existing forms of human factors but for fundamentally new types of human factors. Yet for all the opportunity one has to acknowledge some occasional disappointment. In a very real sense, where *did* the human go? Speaking from a human factors perspective, some of the more military orientated work in this domain seems stronger on doctrine and box diagrams than it does on theory and evidence. Particularly noteworthy are concepts with a long legacy and substantial theoretical underpinning in human factors, re-invented without a great deal of reference to what has gone before, or worse, the human in NEC is reduced to a form of rational optimiser and mathematically modelled from there.

If that is the human factors view, then from the other side of the fence the military reader looking in would probably be inspired and disappointed in equal measure. Common criticisms of human factor's initial forays into NEC are often rooted in what is perceived as a lack of foundational rigour and mathematical profundity, something that can, and routinely is, levelled at all social sciences. This combines with what can often seem like naivety and a lack of appreciation for military context. All told, the human in NEC represents a very difficult interface and it is clear that NEC is uncommonly prone to scientific antagonism of precisely this sort. Quite simply, every time one crosses an interdisciplinary barrier, of which there are many in NEC research, there is the ever present risk of ruffling feathers and seeming to ride roughshod over any number of finely tuned concepts. It would certainly be easy for the work presented in this volume to be construed in exactly the same way, but that would be to mistake its message.

If the nexus of various disciplines impinging on NEC are seen as a form of Venn diagram then we are not necessarily concerned with the larger parts of allied disciplines which do not overlap and for which a human factors approach is not appropriate; instead we are concerned with the smaller areas which *do* overlap. This book, rather like the fundamental systems concepts underlying NEC itself, is not focused merely on parts but on the connections between those parts. In line with this motif we seek to build linkages, however tentative, that extend the reach of human factors towards other domains such as organisational theory, complex systems research and military operations. For the reader approaching the work from any of these individual specialisms, there may not be much that is fundamentally new apart from the way it has been applied. Rather than re-invent the wheel and engage in an antagonistic form of science, quite the opposite is intended. The spirit in which this book is written is for these linkages to represent bridgeheads from which human factors tries to reach out across its various interdisciplinary boundaries in the hope that other disciplines can see enough that is relevant to want to reach over from the other side. The linkages may at times appear tentative, focusing on breadth rather than depth, incomplete, crude even, but no apology is made for trying to establish what constitutes a stating point. It is from such a point that those within the NEC research community, human factors included, can begin to speak to each other. The message is one of reconciliation, of an interconnected scientific approach that could potentially become, like NEC itself aspires, to become more than the sum of its disciplines.

Sociotechnical Theory

John Gartska was interviewed by a journalist in 2007 and rightly defends the NEC concepts he helped to set in motion with his and Cebrowski's paper. Even he, though, acknowledged that in the short space of a decade the net-centric vision has changed:

> 'You have to think differently about people' he was noted as saying. 'You have your social networks and technological networks. You need to have both.' (Shachtman, 2007, p. 8)

This is the precise sentiment behind the work contained in this volume.

In human factors literature (and beyond) the word sociotechnical is ubiquitous. On the one hand you have the socio, of people and society, and on the other the technical, of machines and technology. Socio and technical combine to give 'sociotechnical' (all one word) or 'socio-technical' (with a hyphen). Both variations mean the same thing. In the human factors world we speak freely of 'purposeful interacting socio-technical systems...' (Wilson, 2000, p. 557), 'complex sociotechnical systems ...' (Woo and Vicente, 2003, p. 253), 'sociotechnical work systems...' (Waterson, Older Gray and Clegg, 2002, p. 376) and many more besides. But what does it actually mean?

Like command and control, the phrase 'sociotechnical' has become a synonym, a descriptive label for any practical instantiation of socio and technical, people and technology, the soft sciences meeting hard engineering. In these terms sociotechnical does not mean a great deal as it is difficult to imagine any meaningful system these days that is not described by these two worlds colliding. Whether designed, manufactured, used, maintained or disposed by humans, humans are involved somewhere along the line and the term sociotechnical emerges merely as a convenient, if slightly tautological, buzzword for what results. There is more to the phrase sociotechnical than this, however. In fact, sociotechnical has a very specific meaning.

Sociotechnical 'theory' refers to the interrelatedness of 'social' and 'technical' and is founded on two main principles. One is that the interaction of social and technical factors creates the conditions for successful (or unsuccessful) system performance. These interactions are comprised partly of linear cause and effect relationships, the relationships that are normally designed, and partly from non-linear, complex, even completely unpredictable and chaotic relationships, those that are unexpected.

Whereas the 'technical' is assumed to be the former (i.e., linear), the latter (i.e., non-linear) tends to be the latter domain of the 'socio'; after all, people are not rational optimisers and they do not behave like machines. Then again, as machines and technology venture further along the transformation from platform centric to net-centric, when even simple-minded technology is connected together, what often results is the unpredictable, emergent, non-linear behaviours more associated with the socio than the technical. Inevitably, then, what one usually gets when socio and technical systems are put to work are both types of interaction: linear and non-linear, safe and predictable versus unsafe and unpredictable, effects that are directly proportionate to the sum of their causes combined with other effects that are grossly disproportionate. This is the first principle.

Sociotechnical theory's answer is the second of its two main principles: joint optimisation. The argument is that the optimisation of either socio, or far more commonly, the technical, tends to increase not only the quantity of unpredictable, un-designed, non-linear relationships but specifically those relationships that are injurious to the system's performance. Here we are talking of self-destruction rather than self-synchronisation.

How does one achieve joint optimisation? The further argument put forward by the sociotechnical school of thought is to combine the conceptual goal of sociotechnical theory with systems theory to give 'sociotechnical systems theory' (SST). The term sociotechnical 'system' is also frequently used as a descriptive label, as a synonym for any mixture of people and technology, but like sociotechnical 'theory' it refers to an explicit set of concepts deployed in order to design organisations that exhibit open systems properties, and by virtue of this, cope better with environmental complexity, dynamism, new technology and competition. The main open systems metaphors developed and deployed by SST are described in Chapter 2.

Parallel Universes

Chapter 2 demonstrates that the parallel universes of NEC and sociotechnical systems theory overlap. This is the first contention put forward by this book; that NEC is not alone. Where NEC speaks of peer-to-peer interaction and ad-hoc groups, SST speaks of semi-autonomous groups. Where the former speaks of effects based operations the latter speaks of minimal critical specification, and so it goes on, to such an extent that there is considerable and almost complete overlap. The key difference is that whereas the former is based on around 10 years of military application (with promising but undeniably mixed results) the latter is based on over 50 years of commercial application with an overwhelming tack record of success. It would indeed be 'false pride' to not explore such a symmetry further.

Chapter 3 continues to work on this contention. Contemporaneous with SST's birth in the 1950's is the overlapping theme of social network analysis. NEC is overlain across a classic empirical study into the effects of different network structures, not just in terms of their outright performance but also the experience of those at work within them. Joint optimisation in the classic sociotechnical sense sees the structure of an organisation as a major tool in helping to create the conditions for cohesion, trust, shared awareness and all the other facets of joint optimisation and self-synchronisation. This link is made explicit in Chapter 3 and developed throughout the book.

Matching Approaches to Problems

Chapter 4 interrupts the progression from SST's considerable legacy in the commercial world and the supporting empirical studies contained in Chapter 3 by getting in touch with a concept fundamental to both SST and NEC, that of complexity. The question is directed to exactly what NEC and SST are an organisational response to? Taking a lead from the extant NEC literature the answer is clearly complexity. The difference here is that the concept is considered from a social/human science perspective.

Despite increasing lip service being paid to it in human factors literature, the study of complexity has not generally been followed up with an in-depth exploration of what it actually is and what it means. Chapter 4 seeks to address this and align human factors with NEC in this important respect. It does so by pulling through a number of key concepts from complex systems research and recasting them in a human factors mould. In the course of doing this we put forward a number of arguments to support our second contention, that if complexity is such a good word for describing many human factors problems (not least NEC) then the corresponding human factors approach has to be matched to it in quite a fundamental way.

What is the Network Enabling?

Chapter 5 continues the progression from legacy to live-NEC. Consistent with matching approaches to problems the NATO SAS-050 approach space, a popular and widely accepted model of command and control, is extended through the use of social network analysis and applied in a live context. Reference to social network analysis has already been made in Chapter 3, and it combines with the insights presented in Chapter 4 to inspire the method by which we have been able to turn the NATO SAS-050 model of command and control from a typology into a practical, usable taxonomy.

Numerical metrics can be extracted from large scale case studies, enabling the approach space to be populated with live data and for the organisation's structure to determine where in the approach space it resides. This answers the fundamental question as to what type of organisation NEC actually is. If structure is an important determinate of joint optimisation then it is of further interest to note that, in real life, NEC changes its structure contingent on both function and time. This links to our third contention, which is that the organisational agility and tempo observed during a live case study results in large part because of the humans in the system. Moreover, it was due in similarly large part to the 'unexpected' adaptations they undertook when faced with the initial conditions that the NEC system in this case represented. In other words, the actual region in the approach space adopted by the organisation did not match the expected region. We examine why.

An Interconnected Approach to Analysis

Now armed with live and empirical data from Chapters 3 and 5 respectively, Chapter 6 sets it once again against a complexity related backdrop. Now data driven, further insights can be pulled through and mapped onto human factors. In this context sensitive dependence on initial conditions relates to a shift away from assuming the human system interaction to be stable, to instead assuming to be inherently unstable. The nature of this fundamental instability is considered in turn from the perspective of emergence. Tentative steps are taken towards understanding this concept from a human perspective and leveraging such insights to determine what human factors methods are best suited to understanding it. Next, the link from complex systems research to the NATO SAS-050 approach space is further elaborated, as it is from this that the 'extended' approach space was inspired. By these means the NATO SAS-050 approach space gains a sizeable and important retrospective legacy.

From Theory to Practice

Chapter 7 maintains the level of analysis at that of the total system by continuing to deploy the extended NATO SAS-050 approach space. Having accumulated a body of evidence in favour of a sociotechnical approach to NEC system design, the question arises as to specifically what form should this take? What organisational structures are expressive of joint optimisation as well as representing a match in terms of approach and putative problem? In order to explore this high level question the NATO SAS-050 model is populated with a wide variety of organisations, from commercial concerns through to terrorist networks. This enables the live instance of NEC presented in Chapter 5 to be set into some kind of real-world context.

From this analysis it emerges that many of the organisations which seem to cope with the challenges of the information-age posses the structural properties of a 'small-world' network. To the extent that this configuration is useful, how does one go about designing an organisation to exhibit such properties? The answer, surprisingly, is to design it according to sociotechnical principles. In this we see, once again, demonstrable evidence that the sociotechnical approach overlaps with yet another of the disciplines that impinge on NEC, not just social network analysis, complexity and organisational science, but also cutting-edge research into network topologies. The sociotechnical approach, therefore, is expressive of the book's fourth main contention, which is to advocate (despite ruffled feathers and occasional analytic crudities) an interconnected approach to analysis and to NEC's manifold issues, significant amongst which is of course the human.

In Chapter 8 the perspective inverts and the analysis descends from the top-down view of jointly optimising network structures to the bottom-up view of the individual 'networked interoperable components' that reside within them. With equipment that is journeying along the transition from platform-centric to net-centric the corresponding human factors approach taken towards them requires scrutiny. In this we see evidence of NEC research informing human factors, not merely the other way around. We argue that in the face of such a transition the systems concepts which apply so well at the level of the total system now also apply to the level of the components, or at least those which increasingly exhibit more systemic behaviour. Inspired by the extent of unexpected adaptation in the real-world case of NEC covered in Chapter 5, as well as the extant literature, the final contention is that the goal of human centred design shifts from designing an end product towards the creation of entities and artefacts that encourage favourable adaptations from initial conditions. In other words, equipment needs to be designed so that it is as agile and self-synchronising as the system within which it resides. This in turn requires a new domestication of systems concepts.

Summary

Despite the various arguments, contentions and themes, the aim of this book is an overridingly simple one: to apply SST to the practice of NEC system design. This task is undertaken against a wider backdrop of Human Factors Integration (HFI). If HFI is about '...providing a balanced development of both the technical and human aspects of equipment provision' (MoD, 2000, p. 6) then this book sets out to describe the way in which SST offers a set of theories, empirical evidence, practical measures, and most importantly, a legacy of delivering on such objectives.

Chapter 2
Reconsidering the Network in NEC

Aims of the Chapter

The key issue for traditional hierarchical command and control is that it is increasingly challenged by a host of distinctly modern problems, namely, environmental complexity, dynamism, new technology, and competition that is able to exploit the weaknesses of an organisational paradigm that has been dominant since the Industrial Revolution. The conceptual response to these challenges is NEC. The aim of this chapter is to show that although developed independently, NEC exhibits a high degree of overlap with concepts derived from sociotechnical systems theory (SST). Sociotechnical theory brings to NEC research much more than merely conceptual overlap. It also brings a successful 60-year legacy in the application of open-systems principles to commercial organisations. This track record is something that NEC research currently lacks. This chapter reviews the twin concepts of NEC and sociotechnical systems theory, the underlying motivation behind the adoption of open systems thinking, provides a review of classic sociotechnical studies and the legacy that comes with it. It is argued throughout that 'classic' sociotechnical systems theory has much to offer 'new' command and control paradigms.

Rationality and Industrial Age Thinking

'Formal organisation design, or deliberate as opposed to informal or evolved organisation design, is part of the evolution of both Western and Eastern civilizations' (Davis, 1977, p. 261). Organisations are more or less synonymous with the descriptive use of the term 'sociotechnical system' in that they are invariably a mixture of socio and technical. Not all organisations, however, are sociotechnical systems in the systemic, jointly optimised usage of the term (as is used in this book). To be clear, organisations, of which military command and control is but one example, are entities to which sociotechnical systems theory can be applied.

In virtually all developed civilisations the recent history in organisational design is wedded to a shared 'industrial age' mindset (Beringer, 1986), one that forms the backdrop to both sociotechnical systems and NEC, and one that can be explained with reference to a deeper, more fundamental concept. Formal rationality (Weber, 2007; Ritzer, 1993; Trist, 1978) lends weight to the opinion of many eminent observers over the years who have been exercised not by the apparent mastery of human endeavour over nature but instead the various maladies. None other than Elton Mayo (1949) surveyed the state of affairs yielded by 100 years of industrial activity and wrote, 'To the artist's eye, some-thing was decidedly eschew in the actual Victorian progress; and that something continues to this day' (p. 4). In spite of all the time, effort and expense that feeds into the design of organisations (e.g., Ritzer, 1993; Davis, 1977), systems (e.g., Bar-Yam, 2003), major projects (e.g., Morris and Hough, 1987) consumer products (e.g., Green and Jordan, 1999) and now NEC (e.g., Baxter, 2005), what consistently emerges is something that is often substantially less effective than intended (Clegg, 2000). More than that, these are entities and artefacts that are occasionally injurious to human well being (although technically effective they are often criticized for being 'dehumanizing'; Ritzer, 1993) and may, in extreme cases, become 'anti-human'. In military command and control the organisational aetiology of friendly fire incidents seems to be a case in point.

Formal rationality is a prominent part of an 'implicit theory' that has guided modern organisational design since the Industrial Revolution. A formally rational organisation is labelled a bureaucracy, in the scientific rather than pejorative sense, and the stereotypical case of so-called 'classic' hierarchical command and control (C2) fits into this category well. Rationalising organisations exhibit a tendency towards hierarchies, reductionism and the maximisation of the following attributes:

- Efficiency: A rational organisation is '...the most efficient structure for handling large numbers of tasks...no other structure could handle the massive quantity of work as efficiently' (Ritzer, 1993, p. 20),
- Predictability: 'Outsiders who receive the services the bureaucracies dispense know with a high degree of confidence what they will receive and when they will receive it' (Ritzer, 1993, p. 21),
- Quantification: 'The performance of the incumbents of positions within the bureaucracies is reduced to a series of quantifiable tasks...handling less than that number is unsatisfactory; handling more is viewed as excellence' (Ritzer, 1993, p. 21),
- Control: '...the bureaucracy itself may be seen as one huge nonhuman technology. Its nearly automatic functioning may be seen as an effort to replace human judgement with the dictates of rules, regulations and structures' (Ritzer, 1993, p. 21).

Like all bureaucracies, classic C2 rests on 'tried and true assumptions: that the whole will be equal to the sum of the parts; that the outputs will be proportionate to the inputs; that the results will be the same from one application to the next; and most fundamentally, that there is a repeatable, predictable chain of causes and effects.' (Smith, 2006, p. 40). As a result, one metaphor for a bureaucracy is as a type of 'organisational machine' (Arnold, Cooper and Robertson, 1995). In other words, 'When all the incumbents have, in order, handled the required task, the goal is attained. Furthermore, in handling the task in this way, the bureaucracy has utilised what its past history has shown to be the optimum means to the desired end' (Ritzer, 1993, p. 20). This is the nub of what formal rationality is really all about. In summary, then, organisations designed along bureaucratic lines can be seen as a way of imposing control theoretic behaviour on a large scale, and in so doing, trying to make inputs, processes, outputs, and even humans, behave deterministically.

The core tenets of formal rationality – efficiency, predictability, quantification and control – in turn link to a much more recent model of command and control developed by NATO working group SAS-050 (NATO, 2007). This model provides three major axes (and a three dimensional space) within which various instantiations of command and control can be plotted. The purpose of defining this so-called 'approach space' is to explore alternative paradigms for command and control, ones that are becoming increasingly tractable with the growth in networked technologies. The formally rational instance of classic, hierarchical C2 can be positioned in the NATO SAS-050 model as shown in Figure 2.1. This type of organisation might be characterised by unitary decision rights (in which

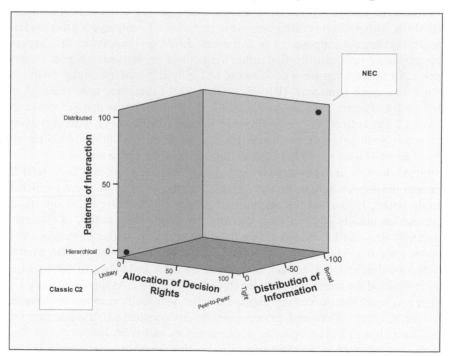

Figure 2.1 **The NATO SAS-050 approach space positions traditional hierarchical command and control in a three-dimensional space defined by unitary decision rights, hierarchical patterns of interaction and tight control of information**

optimum means to ends are specified at the top of, or at higher levels of, a vertical hierarchy); tightly constrained patterns of interaction (owing to rules, standard operating procedures and other means by which the bureaucracy embodies its past experience and specifies optimum means to ends) and tight control (in which performance can be quantified and controlled through intermediate echelons of management). It is these features, and their formally rationalistic backdrop, that together make up the diffuse zeitgeist referred to in contemporary NEC literature as 'industrial age thinking' (e.g., Smith, 2006; Alberts, 2003; Alberts, Garstka and Stein, 1999; Alberts and Hayes, 2006; Alberts, 1996).

The Irrationality of Rationality

The tension created by this prevailing climate of industrial age thinking emerges from '...the observation that Rational systems, contrary to their promise, often end up being quite inefficient' (Ritzer, 1993, p. 122). As Ritzer (1993) goes on to explain: 'Instead of remaining efficient, bureaucracies can degenerate into inefficiency as a result of "red tape" and the other pathologies we usually associate with them. Bureaucracies often become unpredictable as employees grow unclear about what they are supposed to do and clients do not get the services they expect. The emphasis on quantification often leads to large amounts of poor-quality work...All in all, what were designed to be highly Rational operations often end up growing quite irrational' (Ritzer, 1993, p. 22). Experience over centuries of conflict (e.g., Regan, 1991) make it possible to go further to say that in some cases classic C2 can actively create inefficiency (instead of efficiency), unpredictability (instead of predictability), incalculability (instead of calculability) and a complete loss of control (Ritzer, 1993; Trist and Bamforth, 1951). These are the antithetical problems, ironies and productivity paradoxes that, when all else fails, tend to fall into the lap of human factors. The overarching source of these problems, simply stated, is that despite attempts to impose deterministic behaviour on an organisation and its environment, the resultant interlinked entities, or 'system', is actually very hard to maintain in a fixed state (or as something analogous to a 'closed system'). All the more so when such systems are subject to a large range of external disturbances. In fact, the greater the scale and extent of determinism that is imposed the worse the problem can actually become. As these systems grow larger and more interlinked the bigger the effect that such organisations have on their environment. They end up *becoming* the environment and a prime cause of non-linear change and complexity within it (Emery and Trist, 1965).

The evolution of military command and control is particularly instructive. The apotheosis of classic C2 was seemingly reached in the large scale techno-centric style of attrition seen in the cold war era (Smith, 2006). Technically effective hardly seems an adequate term for the sheer destructive might of the opposing military forces in question. Their evolution, for many lesser organisations with militaristic ambitions, served to fundamentally change the environment within which such activities took place. A new paradigm emerged: asymmetric warfare. Asymmetric warfare is characterised by largely urban operations, an opposing force that co-exists with a civilian populous, one that does not adhere to 'rules of engagement' of the sort that classic C2 is adapted. Ironically, the type of organisation that classic C2 now faces is altogether more swarm-like, agile and self-synchronising. It is an organisation that exhibits open systems properties to a far greater degree than the highly technically effective organisation to which these activities are often directed. Is the current situation a case of closed versus open systems? Perhaps. Either way, military organisations around the world, to paraphrase the sociotechnical pioneers Trist and Bamforth (1951), 'seem ready to question a method which they have previously taken for granted'.

Network Enabled Capability

Sitter, Hertog and Dankbaar (1997) offer two solutions for organisations confronted with such difficulties. With an environment of increased (and increasing) complexity, 'The first option is to restore the fit with the external complexity by an increasing internal complexity. [...] This usually means the creation of more staff functions or the enlargement of staff-functions and/or the investment in vertical information systems' (p. 498). And the second option? '...the organisation tries to deal with the external complexity by "reducing" the internal control and coordination needs. [...] This option might be called the strategy of "simple organisations and complex jobs".' This provides a neat characterisation for the current state of affairs in closed versus open systems or asymmetric warfare. One way of putting it is that a traditional military organisation, a complex organisation with simple jobs, is facing a simple organisation with complex jobs.

The purist vision of NEC is congruent with the latter option. Referring back to the NATO SAS-050 model of command and control presented earlier, it can be seen that NEC, unlike classic C2, is characterised by broad dissemination of information and shared awareness. This 'includes the sharing not only of information on the immediate environment and intentions of our own enemy and neutral forces, but also the development of shared combat intent and understanding' (Ferbrache, 2005, p. 104). NEC is further characterised by distributed patterns of interaction and agility. This is the 'ability to reconfigure forces and structures rapidly, building on this shared awareness, exploiting effective mission planning methods, and enabled by an information environment that allows rapid reconfiguration of the underlying network and knowledge bases' (Ferbrache, 2005, p. 104). NEC is also characterised by peer-to-peer interaction and synchronisation, which is the 'ability to plan for and execute a campaign in which we can ensure all elements of the force work together to maximum military effect by synchronizing the execution of their missions to mass forces or generate coordinated effects on target' (Ferbrache, 2005, p. 104). Although there is little evidence of overt cross-referencing, this vision is shared almost exactly with the 'simple organisations and complex jobs' concept, in other words, those organisations designed according to sociotechnical principles.

In the remainder of the chapter our primary task will be to explore the domain of sociotechnical systems theory (SST) and to review exactly how it could bestow NEC-like open systems behaviour upon military command and control. This task will be conducted against a backdrop of continually trying to establish synergies between these two seemingly compatible domains.

Main Principles of Sociotechnical Theory

Sociotechnical theory offers a theoretical basis from which to design organisations, and moreover, to harness the advantages that NEC-like command and control promises. This section goes back to first principles and the seminal work of Trist and Bamforth (1951) entitled, 'Some social and psychological consequences of the longwall method of coal getting'. Despite the coal mining subject matter this is a fascinating and surprisingly relevant case study.

The analysis was motivated by the following irrationality: 'Faced with low productivity despite improved equipment, and with drift from the pits despite both higher wages and better amenities [...] a point seems to have been reached where the [coal] industry is in a mood to question a method it has taken for granted' (Trist and Bamforth, 1951, p. 5).

The longwall method of coal mining represented something akin to mass production. Large-scale machinery was able to cut large swathes of material from the face and represented a considerable economy in man power and the potential for high productivity. At a more fundamental level this method of working also reflected a number of rationalistic principles:

- Large-scale coal cutting machinery led to a simplification and specialisation of the miners tasks.
- The method of working became driven by the needs of the mechanised method with the pattern of the shift and its social organisation changing as a result.
- This new organisation required an intermediate level of supervision and management that was previously absent.

The NATO SAS-050 approach space would characterise the longwall method of coal getting by its hierarchical patterns of interaction (the large scale operations required the task to be decomposed...), its unitary allocation of decision rights (...with one person in charge of the shift...) and its relatively tight control over the distribution of information (...due to increasing role specialisation). All of this means that it is conceptually very similar to classic C2 despite the vast differences in task type and domain. In sociotechnical terms, the new mechanised longwall method of mining represented something of a retrograde step compared to the previous 'hand got method'.

The hand got method divided up the coal getting task along the coal face, with miners working in closely knit teams, mining a certain section of the face from start to finish with all members of the team able to undertake all parts of the task. This method, despite its arcane outward appearance, was nevertheless characterised by a broader dissemination of information (all team members knew what was expected of them…), more distributed patterns of interaction (…the structure of the shift was less hierarchical) and more devolved decision rights (…with greater autonomy bestowed upon the team). Trist and Bamforth explain that, 'the longwall method [can be] regarded as a technological system expressive of the prevailing outlook of mass-production engineering and as a social structure consisting of the occupational roles that have been institutionalized in its use' (1951, p, 5). The prevailing outlook that they refer to is the industrial age rationalizing method of the 'factory system' (the Fordist production line) applied to coal mining. They continue: 'These interactive technological and sociological patterns will be assumed to exist as forces having psychological effects in the life-space of the face worker, who must either take a role and perform a task in the system they compose or abandon his attempt to work at the coal face' (p. 5). The psychological effects of the interacting socio and technical 'forces' (to use the language of 1950s sociotechnical theory) was leading to reduced productivity and increased absenteeism. The notion of 'interactive technological and sociological patterns' quickly evolved to become the term 'sociotechnical'.

In order to explore this productivity paradox 20 miners with varied experience of the work domain were interviewed at length, along with various management and training role incumbents. Trist and Bamforth's paper is thus based on ethnographic techniques, the outcomes of which led to the elaboration of a number of enduring sociotechnical principles which include the following:

Responsible Autonomy

'The outstanding feature of the social pattern with which the pre-mechanized equilibrium was associated is its emphasis on small group organisation at the coal face'. Indeed, 'under these conditions there is no possibility of continuous supervision, in the factory sense, from any individual external to the primary work group' (Trist and Bamforth, 1951, p. 7). Harsh physical constraints quite simply prevented the task from being carried out, with this technology, in any other way. So instead of a larger 'whole of shift' based organisation accountable to intermediate layers of management, the hand-got method embodied internal supervision and leadership at the level of the 'group' which resulted in so-called 'responsible autonomy' (Trist and Bamforth, 1951, p. 6). Sociotechnical theory was pioneering for its focus on the group as the primary unit of analysis.

A facet of this method of working that is somewhat unique to the dangers of the underground situation, yet with ready parallels to military operations, is 'the strong need in the underground worker for a role in a small primary group' (Trist and Bamforth, 1951, p. 7). It is argued that such a need arises in hazardous circumstances, especially in cases where the means for effective communication are limited. As Carvalho (2006) states, operators use this proximity and group membership '...to produce continuous, redundant and recursive interactions to successfully construct and maintain individual and mutual awareness...' (p. 51). The immediacy and proximity of trusted team members makes it possible for this need to be met with favourable consequences for team cohesion, trust and overall system performance. The field of NEC is rightly interested in the varied issues of trust and team cohesion (e.g., Siebold, 2000; Oliver, et al., 2000; Mael and Alderks, 2002) and whilst the principles of sociotechnical systems theory are far from a panacea they do at least admit the possibility of creating favourable conditions for these varied aspects to emerge.

Adaptability

As Trist and Bamforth put it, 'though his equipment was simple, his tasks were multiple', the miner '...had craft pride and artisan independence' (1951, p. 6). The 'hand-got method' is an example of a simple organisation (and equipment) 'doing' complex tasks. The longwall method, on the other hand, is an example of a complex organisation (and technological infrastructure) 'doing' simple tasks. Job simplification has long been associated with lower moral and diminished job satisfaction (e.g., Hackman and Oldman, 1980; Arnold, Cooper and Robertson, 1995). In the former case a type of 'human redundancy' was created (e.g., Clark, 2005) in which 'groups of this kind were free to set their own targets, so that aspiration levels with respect to production could be adjusted to the age and stamina of the individuals concerned' (Trist and Bamforth, 1951, p. 7). This meant that outcomes or 'effects' were more important than activities or the precise means by which those effects were achieved.

Trist and Bamforth (1951) go on to note that 'A very large variety of unfavourable and changing environmental conditions is encountered at the coal-face, many of which are impossible to predict. Others, though predictable, are impossible to alter.' (p. 20). The longwall method was clearly inspired by the appealing industrial age, rational principles of 'factory production' wherein 'a comparatively high degree of control can be exercised over the complex and moving 'figure' of a production sequence, since it is possible to maintain the 'ground' in a comparatively passive and constant state' (Trist and Bamforth, 1951, p. 20). In many contexts, coal mining and military operations being just two, there is relatively little in the way of opportunity for maintaining the 'ground' in such a state. If the environment cannot be made to approximate to the linear cause and effect logic of determinism then 'the applicability [...] of methods derived from the factory' is limited (Trist and Bamforth, 1951, p. 20).

Meaningfulness of Tasks

Sociotechnical theory is as concerned for the experience of humans within systems as it is with the system's ultimate performance. Under the rubric of joint optimisation sociotechnical systems theory sees the two as complementary. Trist and Bamforth (1951) go into detail as to how this complementary relationship was realised in their mining example. They identify the hand-got method as having 'the advantage of placing responsibility for the complete coal-getting task squarely on the shoulders of a single, small, face-to-face group which experiences the entire cycle of operations within the compass of its membership.' Furthermore, 'for each participant the task has total significance and dynamic closure' (Trist and Bamforth, 1951, p. 6). It is a meaningful task. Meaningfulness arises out of a focus on the group, from responsible autonomy and from adaptability, linking jointly optimised sociotechnical systems to a number of 'core job characteristics' (Hackman and Oldman, 1980):

* Skill variety (e.g., simple organisations but complex varied jobs that rely on a multiplicity of skills; Sitter et al., 1997).
* Task Identity (e.g., 'entire cycle of operations' or whole tasks; Trist and Bamforth, 1951).
* Task Significance (e.g., 'dynamic closure' and meaningful tasks; Trist and Bamforth, 1951).
* Autonomy (e.g., human redundancy, adaptability, semi-autonomous work groups; Trist and Bamforth, 1951).
* Feedback (e.g., continuous, redundant and recursive interactions; Carvalho, 2006).

The pioneering work of Trist and Bamforth was motivated by the most prominent irrationality of rationality, namely, that organisations designed to function deterministically like machines are dehumanising. It elaborated on the central themes that would form a fully fledged sociotechnical school of thought, as well as the specific features that would bestow desirable open systems characteristics on organisations which previously behaved in a way analogous to closed systems. There are ready parallels between Trist and Bamforth's foundational work in the Durham coal mines and the dynamic, uncertain and often dangerous world of military operations. Sociotechnical-systems theory provides a detailed conceptual language with direct links to NEC system design. Perhaps the major influence of Trist and Bamforth's pioneering sociotechnical work, however, was to change the prevailing viewpoint in which organisations were considered: from a purely technical perspective (industrial age thinking) or as purely social entities (an organisational or industrial relations perspective) to instead '...relate the social and technological systems together' (Trist, 1978, p. 43).

Sociotechnical Systems

Systems thinking offered the conceptual language from which notions of 'networks' and 'distributed systems', and indeed, NEC itself ultimately derive. Its application to sociotechnical theory came in 1959 with a paper by Emery which expanded the field by drawing on the specific case of open systems theory (Kelly, 1978, p. 1075; Bertalanffy, 1950). Open systems theory gave sociotechnical theory a more tightly defined grounding as well as a unifying conceptual language.

The characteristics of systems thinking are the twin notions of 'a complex whole' formed from a 'set of connected things or parts' (Allen, 1984). Part of the appeal of industrial age thinking is that the 'set of connected things or parts' can be tightly defined. A visual metaphor for such a deterministic system might be an electrical circuit diagram, an artefact with components that have known input/output properties connected by electrical pathways with similarly known properties and flows. Such an artefact, in systems terms, could be referred to as a closed system or an 'object' or a rational system. We of course have to rely on a degree of analogy here. Whilst not seeking to push the analogy too far, we can state that in organisational terms a closed system can be said to be concerned with the attainment of a specific goal and there is a high degree of formalisation (Scott, 1992). Another word used to describe such an organisation is a bureaucracy. An open system is quite different.

Import and Export

Open systems are acknowledged to have boundaries with other systems and some form of exchange existing between them: 'A system is closed if no material enters or leaves it; it is open if there is import and export and, therefore, change of the components' (Bertalanffy, 1950, p. 23). In the original biological conception of open systems this exchange would be 'matter' such as haemoglobin or oxygen. As systems theory has expanded, the inviolable characteristic of all such exchanges is now seen as essentially informational (e.g., Kelly, 1994; Ciborra, Migliarese and Romano, 1984). Exchange between system elements is input, which causes state changes, outputs of which become further inputs for other elements. An appropriate visual metaphor for such a system might be a block or Venn diagram in which the properties of the components and the links between them are not as well defined as an electrical circuit diagram and are subject to change. A system exhibiting these properties is also referred to as a network, expanding somewhat the definition of 'network' in NEC from that of a computer network to a formal systems metaphor.

Steady States

'A closed system must, according to the second law of thermodynamics, eventually attain a time-independent equilibrium state, with maximum entropy and minimum free energy [...] An open system may attain (certain conditions presupposed) a time-independent state where the system remains constant as a whole... though there is a constant flow of the component materials. This is called a steady state' (Bertalanffy, 1950, p. 23). If sociotechnical systems are open systems then organisations become analogous to a 'vitalistic' (i.e., living) entity. The idea of a 'vitalistic entity' strikes a chord in organisational psychology. At least one metaphor for an organisation is 'organismic', meaning that it is able to adapt and evolve to environmental changes and behave more like an ecology than a machine (e.g., Morgan, 1986). This is something that NEC, with its terminology couched very much in terms of 'agility' and 'self-synchronization' is undoubtedly, if sometimes implicitly, striving for. By comparison, a closed system is, or becomes, locked or frozen in a particular state and requires no further import or export of information to maintain that state. A closed system (or industrial age organisation taken to its extreme) is therefore unresponsive to environmental change, matched to an optimum means to an end within a defined context and slow to change or adapt, if at all. To use a computer science metaphor, a closed system could be described as an entity that is 'programmed' while an open system is an entity that can 'learn'.

Equifinality

Related to these ideas of dynamism and adaptability is the notion of equifinality, described by Von Bertalanffy thus: 'A profound difference between most inanimate and living systems can be expressed by the concept of equifinality. In most physical systems, the final state is determined by the initial conditions... Vital phenomena show a different behaviour. Here, to a wide extent, the final state may be reached from different initial conditions and in different ways' (p. 25). This is exactly what NEC desires when it speaks of self-synchronisation and what sociotechnical theory offers in terms of adaptability and semi-autonomy. Equifinality grants open systems a certain 'paradoxical behaviour', 'as if the system "knew" of the final state which it has to attain in the future' (p. 25), which of course it does not, merely that sociotechnical principles permit it to rapidly adapt and evolve to one. Trist (1978) could have been describing NEC when saying that open systems grow 'by processes of internal elaboration. They manage to achieve a steady state while doing work. They achieve a quasi-stationary equilibrium in which the enterprise as a whole remains constant, with a continuous "throughput", despite a considerable range of external changes.' (p. 45). To sum up, then, sociotechnical systems theory is inextricably linked to ideas about open systems. The principles first elaborated by Trist and Bamforth (1951) are framed in terms of endowing favourable open systems behaviour on organisations. Military command and control is one such organisation.

Real-Life Sociotechnical Systems

Classic Examples

Since the pioneering work of Trist and Bamforth in 1951 there has been considerable effort undertaken in organisational design using sociotechnical principles. During the late sixties and seventies, riding on a groundswell of left-leaning ideologies, sociotechnical systems theory was attractive. Indeed, it was the darling of business process engineering occupying a niche which today might be occupied by TQM and Six Sigma. A considerable number of organisations embraced sociotechnical systems methods, the two most famous being Volvo and Olivetti. We begin this section by providing a case study from the 'classic' sociotechnical school. The intention is to provide a grounding and contextualisation, describing what an organisation redesigned according to sociotechnical principles actually 'looks like', and importantly, what it achieves when subject to real-life commercial pressures. Given the synergies already alluded to, the ambitions of NEC would seem to find a ready parallel in the apparent successes of 'real-life' sociotechnical interventions.

It is to one of the earliest accounts of a comprehensive organisational redesign according to sociotechnical principles, undertaken by Rice (1958) in textile mills in Ahmadabad, India, that we now turn. Here, as elsewhere, the sociotechnical redesign led to a radically different organisation which, it was argued, was jointly optimised. More than that, the 'reorganisation was reflected in a significant and sustained improvement in mean percentage efficiency and a decrease in mean percentage damage [to goods]...the improvements were consistently maintained through-out a long period of follow up' (Trist, 1978, p. 53).

As mentioned above, the most famous example of sociotechnical design is undoubtedly that undertaken at Volvo's Kalmar and Uddevalla car plants (e.g., Hammerström and Lansbury, 1991; Knights and McCabe, 2000; Sandberg, 1995). Whilst many commercial instantiations of sociotechnical systems theory are criticised for their limited degree of 'technological' change (choosing to focus instead on the altogether less expensive aspects of 'socio' and 'organisational' change; Pasmore et al., 1982) Volvo appeared to embrace the principles in an uncommonly comprehensive manner, and on a scale heretofore not yet experienced. The defining feature of the Kalmar plant's design was a shift from a rationalistic style of hierarchical organisation to one based on smaller groups. Once again, although a world away from military command and control, conceptually, this shift retains its familiarity. In Volvo's case the change was radical. The production line, the overtly 'technical' in the sociotechnical equation, not to mention the emblem of industrial age organisational design itself, literally disappeared. It was replaced by autonomous group-work undertaken by well-qualified personnel, 'advanced automation in the handling of production material; co-determination in the planning and a minimum of levels in the organisation' (Sandberg, 1995, p. 147). In other words, a team of skilled workers undertook the final assembly of an entire car, from start to finish, using a kit of parts provided by the 'advanced production material automation'. The team could also decide amongst themselves how that task was to be undertaken.

From a systems perspective, according to Dekker's more contemporary work on network structures (e.g., 2002), this new configuration had something of a 'hybrid' feel to it. In structural terms there was a mixture of hierarchical subdivision (albeit to a far lesser extent than before) and peer to peer interaction (within groups rather than everybody literally interacting with everyone else). In the language of NEC research what you had, in effect, were 'communities of interest', shortcuts between these communities and interactions that were 'driven by circumstances' (e.g., Alberts and Hayes, 2005; Granovetter, 1973; Watts and Strogatz, 1998). What we term here as a 'hybrid structure' in fact possesses some quite unique properties which we will explore in much more detail later in Chapter 7. For the time being it can be noted that, 'the learning in this work organisation is impressive. Being engaged in all aspects of work makes the production comprehensible and the employees become, as part of their job, involved in the customer's demands and in striving after constant improvement. Work intensity is high' (Sandberg, 1995, p. 148). Another major effect of this network structure, as Trist (1978) notes, is that 'whereas the former organisation had been maintained in a steady state only by the constant and arduous efforts of management, the new one proved to be inherently stable and self correcting' (p. 53). To put this into the language of systems theory, the organisation started to behave like an open system, one that could achieve a steady state based on a constant throughput of inputs and outputs, and able to maintain this steady state despite considerable changes in the environment. In the language of NEC this phenomenon would be referred to as something like 'self-synchronization' (e.g., Ferbrache, 2005). Whichever language it is translated into, these findings provide practical, large-scale, industrial validation of a conceptual approach to dealing with complexity.

Organisational Interventions

The specific mechanisms instantiated to support this open systems behaviour are varied and interconnected. Principle among them are natural task groupings that bestow a form of autonomous responsibility on the group, there was a 'whole task' to be undertaken and the requisite skills within the group to undertake it from beginning to end. In Volvo's case the parts for the cars were organised as if they were kits, with each member of the team completing a proportion of the kit and the team as a whole effectively building a whole car (more or less independently of other teams). In terms of agility it was quickly observed that 'model changes

[...] needed less time and less costs in tools and training' compared to a similar plant that was organised around the traditional factory principle (Sandberg, 1995, p. 149).

Obviously, such teams still needed to be related to the wider system, which required someone to work at the system boundaries in order to 'perceive what is needed of him and to take appropriate measures' (Teram, 1991, p. 53). In command and control terms this new organisation shifts the primary task of commanders (or managers) away from processes of internal regulation to instead being more outwardly focused (Trist, 1978). Alberts and Hayes in their book 'Power to the Edge' describe a very similar role under the heading of a 'Strategic Corporal'. This is someone who 'must be able to function across a range of missions and make decisions that have implications far beyond his local responsibilities (Alberts and Hayes, 2005, p. 65). Back at Volvo, the similar post was labelled 'lagombud' or 'group ombudsman'. Not only had the assembly line disappeared but so to had the role of supervisor. Instead, here was a strategic corporal 'who relates to other groups and to the product shop manager' (Sandberg, 1995, p. 148). This is an important conceptual difference. Under this sociotechnical paradigm managers and commanders now become a form of executive, coordinating function, 'designing behaviours' rather than arduously 'scripting tasks' (e.g., Reynolds, 1987).

The Sociotechnical Legacy

In order to provide a wider characterisation of the extant work in 'classic' sociotechnical systems theory a number of large meta-analyses have been identified. The first analysis is contained in a paper by Cummings, Molloy and Glen (1977) which reviewed 58 studies, the second in a paper by Pasmore et al., (1982) which reviewed 134 studies, and the third is by Beekun (1989) covering a further 17. Between them they provide a substantial overview of the first 38 years worth of practical experience in this domain.

If this large body of work can be characterised at all then it can be done so with reference to the overwhelming predominance of positive study outcomes. There is no doubt that the combined results 'support most of the claims that [sociotechnical] researchers have been making for three decades concerning the beneficial nature of this organisational redesign strategy' (Beekun, 1989, p. 893). Table 2.1 is synthesised from Pasmore et al., (1982) and provides a good overview not just of the wider range of sociotechnical measures available and implemented in practice, but also an indication of how successful they were in relation to two dependent variables: attitudes (a more 'socio' variable which reflects the experience of those at work within such organisations) and productivity (a variable which reflects a form of 'technical' optimisation).

The positive outcomes derived from applying the sociotechnical interventions listed in the left-hand column of Table 2.1 are striking. It is also interesting to note that by far the most common intervention is represented by structural change and a

Table 2.1   The use and effectiveness of common sociotechnical measures

Feature	Percentage of studies using feature (N=134)	Percentage of studies reporting success in terms of attitudes (Socio)	Percentage of studies reporting success in terms of productivity (technical)
Autonomous work groups	53	100	89
Skill development	40	94	91
Action group	22	93	100
Reward system	21	95	95
Self inspection	16	100	90
Technological change	16	92	60
Team approach	16	100	80
Facilitative leadership	14	100	100
Operators perform maintenance	12	100	88
Minimum critical specification	9	100	100
Feedback on performance	9	100	100
Customer interface	9	100	100
Self-supply	8	100	80
Managerial information for operators	7	100	67
Selection of peers	6	100	100
Status equalisation	4	100	100
Pay for learning	4	100	100
Peer review	3	100	100
% studies successful		94	87

Source: Pasmore et al., 1982.

work organisation based around teams. In other words, adapting the human to the system is far more common than adapting the technical to the human (53 per cent of studies for the former compared to only 16 per cent for the latter).

Table 2.1 shows that not all applications of sociotechnical interventions are successful and it is right to guard against any over-enthusiasm. Indeed, it is certainly the case that the number of unsuccessful attempts that crop up in the public domain are likely to be fewer than those which are successful. Despite that, the preponderance of positive outcomes is difficult to ignore and there is certainly no similar legacy in NEC research. The only downside is that with so little in the way of variance it becomes difficult to judge the effect of any specific sociotechnical intervention (Cummings et al., 1977). Herein lies an interesting point. In systems-thinking it is not generally possible, or even desirable, to trace a generalised effect (productivity or attitudes for example) to specific causes. The point seems to be that sociotechnical principles and interventions are as systemic and equifinal as the system to which they are applied (e.g., Clegg, 2000). Like all good systems, perhaps they, too, become more than the sum of their parts?

From the literature, then, it would seem that implementing an ostensibly 'technical' system like NEC is on a scale considerably in excess of many study domains analysed previously, at least within the so-called 'classic' sociotechnical school. What singles out NEC as a somewhat unique case is its distributed nature. For example, the idea of a roving ombudsman figure is perhaps something of an anathema in cases where teams are distributed nationally and even internationally. This kind of distribution, the separation of information from physical artefacts and locations, is an inherent part of the information-age itself. Some of the lessons to be learnt from the commercial arena will, of course, require further work in order to realise an equivalent in the domain of NEC. Another factor unique to NEC is the degree of non-linearity and complexity inherent in it. Despite the open systems principles created by sociotechnical systems theory the vast majority of the application domains are considerably more deterministic than the military arena, which has the unique property of entities in the environment that are not just dynamic but deliberately and adaptively trying to thwart your activities. There is a need to draw inspiration not only from successful classic sociotechnical studies but to also examine more contemporary developments which seek to move sociotechnical systems theory from its roots in the industrial age to a new information-age context.

Contemporary Sociotechnical Systems Theory

Faced with problems that are increasingly framed in terms of non-linearity and/or complexity, approaches such as macro ergonomics (e.g., Kleiner, 2006; Kirwan, 2000), cognitive systems engineering (e.g., Hollnagel, 1993; Hollnagel and Woods, 2005) and other nascent systems based developments attest to a growing shift in human factors methods and modes of thought. As for organisations, the shift here was not due to the inherent lack of humanism but simply their 'inability to adapt to rapid change' (Toffler, 1981, p. 135/6). As for SST itself, British readers with a passing knowledge of industrial relations history will probably question the paradox that in the 1970s the National Union of Mineworkers voted for strike action. Their primary concern was pay but it was also to do with the hand-got method described above by Trist and Bamforth. One of the major complaints was that the working conditions were 'un-mechanized' (Turner, 2008). Moving forward in time to November 1992 and what most traditional texts on organisational design will tend not to mention is that the most famous sociotechnical case study of all, the Volvo Kalmar plant, is now closed.

This story of apparent rise and fall might seem to cast sociotechnical systems in a poor light, as representing a slightly Luddite approach to systems design, one that has had its day and is no longer relevant to modern concerns. This would be a mistake. In the case of the coal-mining example, the sociotechnical approach did not advocate the primitive hand-got method as such; rather, it recognized that the work organisation had resulted from the co-evolution of people and environment and that large scale mechanisation represented a step change which did not necessarily recognise the favourable aspects of this hard-won experience. Sociotechnical systems are not anti-technology. As for the Volvo Kalmar plant, the reasons for its closure were considerably more obtuse. On balance, the closure of the plant was *not* due to a failure of the sociotechnical paradigm. Instead, it was more to do with the same shifting paradigm which favoured NEC in its early vertically integrated shock and awe form, and the resurgence of neo-Taylorism inspired by the manufacturing excellence then evident in Japan (Dankbaar, 1993). This shift coincides with the so-called 'information revolution' and the increasingly inventive ways in which computing and networked technology could be deployed to make complex organisations behave rationally in the face of complexity. There is no doubt that the subsequent character of sociotechnical research has been affected. Certainly, the days of ambitious large scale implementations of sociotechnical principles have largely given way to work of a smaller and somewhat more self-effacing nature, some of which is surveyed below. Current sociotechnical thinking does, however, share with NEC an interest in the opportunities and issues raised by information technology and the internet. SST is itself shifting as evidenced by work of a more contemporary nature.

Consider that a central focus of modern organisations, NEC included, is to be highly responsive to the needs of the recipients of the services which the organisation dispenses; that is to say organisations should be able to learn, the quicker and more adaptively the better (Adler and Docherty, 1998). Hirschhorn et al. (2001) refer to the idea as 'mass-customization' in which the real value of joint optimisation is not in the production and dissemination of 'things' (e.g., physical goods or 'actions') but in the production and dissemination of information (e.g., informational commodities like 'effects'). They present the data shown in Table 2.2 to illustrate the conceptual shift.

Table 2.2   Comparison of sociotechnical contexts

Focus of sociotechnical systems is on:	Focus of NEC is on:
Mass production	Mass customisation
Minimising down time	Minimising learning time
Producing product	Producing information
Maintaining a steady state	Finding information
Performing work sequentially during a defined 'run'	Performing work continuously and adaptively

Source: Adapted from Hirschhorn et al., 2001, p. 249.

Table 2.2 clearly shows that an emphasis on traditional contexts and the micro level of organisational design would diminish the effectiveness of SST in modern settings, hence the currently expanding array of contemporary sociotechnical concepts (e.g., Heller, 1997, p. 606). Examples of these include 'open sociotechnical systems' (e.g., Beuscart, 2005) and sociotechnical capital (e.g., Kazmar, 2006) to name but two. The term open sociotechnical systems is an apparent contradiction in terms but actually refers to the nature of the group and to its flexible open membership. The shift is from relatively enduring semi-autonomous groups towards comparatively transient ad hoc groups. Sociotechnical capital is drawn from research in the world of the internet and a growing fascination with emergent phenomena that arises from 'mass collaboration' (e.g., Tapscott and Williams, 2007) and online communities. Sociotechnical capital deals with the formation and regulation of such groups and the characteristics of network systems required to support them (e.g., Resnick, 2002). Scacchi (2004) provides a contemporary summary of future research directions which, if anything, seem to be aligning the world of sociotechnical systems theory ever more closely with the concerns of NEC. They are as follows:

- 'First, the focus of [SST] design research is evolving towards attention to [sociotechnical interaction networks (STINs)] of people, resources, organisational policies, and institutional rules that embed and surround an information system' (p. 5). The question of where socio and technical boundaries lie is becoming ambiguous, as is 'who' the user of a system is (and the panoply of values, perspectives, demands etc., that they require and/or impose).
- Second is 'recognition that a large set of information systems in complex organisational settings generally have user requirements that are situated in space (organisational, resource configuration, markets, and social worlds) and time (immediate, near-term, and long term), meaning that user requirements are continuously evolving, rather than fixed' (p. 6).

This prompts a need to decide how to access and respond to these shifting informational requirements and is reflection of similar powerful trends in Systems Engineering.

- Third, it is unclear how best to 'visualize, represent, or depict (via text, graphics, or multi-media) a sociotechnical system' (p. 6). So-called 'Rich Pictures' (e.g., Monk and Howard, 1998), social network diagrams (e.g., Driskell and Mullen, 2005), soft systems methodologies (e.g., Checkland and Poulter, 2006) and the Event Analysis for Systemic Teamwork methodology (e.g., Stanton, Baber and Harris, 2007; Walker et al., 2006) already reflect work underway to try and address this issue. Later chapters contribute even more directly to this aim.

- Fourth, 'the practice of the design of sociotechnical systems will evolve away from prescriptive remedies to embodied and collective work practices' (p. 7). An example is given of 'free/open source software development projects or communities. In this sociotechnical world, the boundary between software system developers and users is highly blurred, highly permeable, or non-existent' (p. 6–7). For example, in adapting and modifying NEC to suit local needs and preferences (as is inevitably the case; e.g., Verrall, 2006) users of NEC play as much a part in the design of the final system/ organisation as the designers themselves; whether they like it or not.

Clearly this is not an exhaustive list but it is, we believe, a characterisation, one that is expressive of the changing nature of SST, the changing nature of the systems (and their boundaries) and the changing nature of the environment to which they are applied.

Summary

Despite its ubiquity within human factors literature the term 'sociotechnical' is clearly much more than merely a buzzword. It is a set of explicit concepts, inspired by general systems theory, aimed at jointly optimising people, technology, organisations and all manner of other systemic elements. This introductory review has highlighted a key set of basic sociotechnical principles (responsible autonomy, adaptability and meaningfulness of tasks) which seem to offer favourable initial conditions for effective NEC systems. Sociotechnical principles create shared awareness (through peer to peer interaction) and agility (through effects based operations, semi-autonomous groups and increased tempo) and self-synchronisation (by joint optimisation and synergy). As Table 2.3 illustrates, the extent of overlap between the classic concept of SST and the new command and control paradigm heralded by NEC is clearly manifest.

The sociotechnical approach challenges the dominant techno-centric viewpoint whereby the 'Network' in NEC is seen in terms of merely data and communication networks. Instead, SST speaks towards the optimum blend of social *and* technical

Table 2.3   Comparison of concepts: NEC versus Sociotechnical Systems

NEC concepts	Sociotechnical concepts
Agility and tempo	Adaptability
Effects-based operations	Minimal critical specification
Peer-to-peer interaction and ad-hoc groups	Semi-autonomous work groups
Self-synchronisation	All of the above combined with whole and meaningful tasks

networks. It also refines the notions of shared awareness (the extent to which everybody needs to know everything), peer-to-peer interaction (semi-autonomous groups are a particular form of this) and, at a more fundamental level, represents a conceptual response for dealing with complexity (which is shifted from a global level of complexity to a local view of complexity which semi-autonomous groups respond to faster and more adaptively). Perhaps above all the sociotechnical approach is innately human-centred. It is concerned as much with the optimisation of effectiveness as it is with the experiences of people working within command and control organisations. Indeed, the performance of the system and the experience of those at work within it are mutually reinforcing and complementary. Arguably, the most adaptable components of all within NEC are the human actors. Sociotechnical theory, therefore, brings with it a humanistic value base and set of non-Tayloristic assumptions. It is not being offered as a panacea, but SST seems to offer considerable promise in terms of at least creating the conditions for cohesive, expert, flexible teams that relate well to a wider system. All of this would be mere conjecture were it not for SST's 60 year legacy of applying open systems principles to commercial organisations. Whether, and by what means, the same positive outcomes could be realised in the field of NEC is something that the following chapters go on to deal with.

Chapter 3

Some Effects of Certain Communications Patterns on Group Performance

Aims of the Chapter

The sociotechnical perspective elucidated in the previous chapter naturally obliges us to recognise the role of humans in NEC, yet ironically this is often precisely the aspect omitted, or at least grossly over-simplified, in the design of such systems (e.g., Stanton et al., 2009). All too often, what seems to have occurred is that computer models and simulations have replaced the lab rats that researchers used to use in order to try and understand human behaviour. Harold J. Leavitt's 1951 paper 'Some effects of certain communication patterns on group performance' provided a salutary lesson on the effect that different organisational structures have on human role incumbents – moreover, that it is the humans in the system who bestow upon it the agility, tempo and self-synchronisation that is desired. From this and other work in social network theory has sprung a considerable body of research which today is routinely applied to the problems of NEC. From a sociotechnical perspective, however, there is still a gap. In this chapter we present the results of a replication of Leavitt's pioneering study, updated and refreshed for the information-age and related to the specific concerns of NEC. The task here is not to convince anyone of the validity or legacy attached to Leavitt's foundational work (for an overview of where it fits into the wider pantheon of social network research see, for example, Martino and Spoto; 2006); instead, the goal is to put the human back into NEC and to show, via an experimental demonstration, that the principles of sociotechnical systems and of joint optimisation still apply. Whereas the previous chapter attempted to build a conceptual bridgehead between the world of sociotechnical systems theory and NEC, this chapter attempts to build an experimental one.

Introduction

Background and Context

So what actually happens when humans are set to work within the set of organisational boundaries and constraints represented by NEC? Some insight into this question is provided by looking into a live-NEC field trial where the resultant behaviour of the system was not entirely what was expected (Verall, 2006). A substantial and wide-ranging human-factors analysis revealed that this particular socio-technical system, over time, became progressively more preoccupied with the 'means' to achieve a given end rather than the massing of objectives or 'end states', which is what is normally expected from NEC; indeed, what was expected in this case. Similarly, despite the provision of a networked information infrastructure, individuals and teams either used it in unpredictable ways or else adopted more familiar and presumably easier methods of working. To paraphrase Clegg (2000), what was witnessed here was 'people interpreting the system, amending it, massaging it and making such adjustments as they saw fit and/or were able to undertake' (p. 467). This particular case study example was ultimately successful in what it was trying to achieve which, in turn, creates a different interpretation of the results. In other words, is this apparent *lack* of control necessarily a bad thing?

Drawing from the world whence NEC is often assumed to have come, namely the internet and other similar networked technologies, it is possible to discern several powerful trends in which this form of human adaptability, far from being commanded and controlled out of existence, is instead actively exploited. From the sublime (e.g., the Human Genome Project) to the ridiculous (e.g., Wikipedia), these are both networked, highly distributed, systems embodying the diffuse non-linear causality of peers influencing peers (Kelly, 1994; Tapscott and Williams, 2006; Viegas, Wattenberg and McKeon, 2007). These are entities where the boundary between designers and users has become 'highly blurred, highly permeable, or non-existent' (Scacchi, 2004, p. 6–7). Furthermore, under these 'initial conditions' highly effective and agile forms of organisational infrastructure 'emerge' rather than are created. To use Toffler's (1981) or Tapscott and William's (2007) phraseology, the participants in the observed NEC case study behaved rather like 'prosumers', individuals who see 'the ability to adapt, massage, cajole and generally "hack" a new technology as a birthright' (p. 32). In the human factors world Shorrock and Straeter (2006) remind us that people are still needed in complex command and control systems for precisely this reason. Moreover, this sort of human adaptability is inevitable. Therefore, perhaps a more useful way to look at NEC is not to see it as an end product or an entity that 'is' something, but rather as a process, something that 'becomes' (e.g., Houghton et al., 2006). It seems possible to go even further, to argue that an alternative conception of NEC is not something that can be called a finished article but rather as the set of initial conditions from which the most adaptable component of all, the humans in the system, create the end product most useful for their particular set of circumstances. In this we see clearly why Leavitt and others have placed such an emphasis on humans within organisations rather than lab rats within experiments or simulated agents within computer models (all of which have their place, of course).

The Leavitt Study

Leavitt's 1951 study took this simple expedient, that organisational structure will in part determine human behaviour, and subjected it to a direct empirical test. Leavitt derived four distinct organisational structures, or network archetypes, based on small networks (of five agents). Leavitt stated that if 'two patterns [or structures] cannot be "bent" into the same shape without breaking a link, they are different patterns' (1951, p. 39). What emerged were the 'circle', 'chain', 'y' and 'star' networks shown in Figure 3.1.

In Leavitt's paper, the networks were populated with five subjects who were furnished with a simple problem solving task in a paradigm very similar, incidentally, to Moffat and Manso's ELICIT concept (2008; Leweling and Nissen, 2007; Ruddy, 2007, etc.). The subjects sat in adjoining cubicles with the type of organisational structure (henceforth referred to as a 'network') being instigated through the provision of various open and closed letterboxes in the dividing walls. The subjects each had a set of symbols and the task was to communicate with written notes, passed through the letterboxes, what the symbol common to all subjects was. One hundred people took part in the study, split into twenty groups of five. Each group was subject to fifteen trials on each of the four network archetypes. A wide range of measures were extracted enabling both the socio and technical aspects of performance associated with each network to be derived. A clear performance dichotomy was observed:

> 'the circle [network], one extreme, is active, leaderless, unorganized, erratic, and yet is enjoyed by its members. The [star network], at the other extreme, is less active, has a distinct leader, is well and stably organized, is less erratic, and yet is unsatisfying to most of its members.' (p. 46)

Clearly, then, 'certain communication patterns' did indeed yield 'some effects on group performance'. These effects prove to be of a systematic nature when a

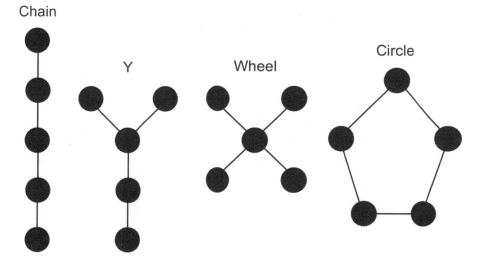

Figure 3.1   Leavitt's network archetypes (1951, p. 39)

further categorisation is applied. Leavitt's Star, Chain and Y networks map, to some extent at least, onto notions of 'classic' hierarchical command and control. The Circle network, on the other hand, is conceptually closer to NEC. Thus in broad terms it can be seen that NEC leans more towards the 'active, leaderless, satisfying' and jointly optimised end of the performance spectrum compared to the technically optimised 'stable, organised and unsatisfying' characteristics of classic C2. This is a very crude analogy but it is not as tenuous as it might appear. Chapter 7 shows that this mapping is in fact well founded. Table 3.1, in the meantime, provides a more detailed comparison of the results obtained.

Table 3.1   Performance characteristics of Leavitt's (1951) network archetypes. Shading denotes best performance in that category

	Network	Overall	Fastest trial	Errors $(\bar{X})$	N of comms. $(\bar{X})$	Cohesion/ enjoyment
NEC	Circle	'showed no consistent operational organisation' p. 42.	50.4s	3.4	419	82/100
Classic C2	Y	'operated so as to give the most central position [...] complete decision-making authority. [...] Organisation for the Y evolved a little more slowly than for the [star], but, once achieved, it was just as stable' p. 42.	35.4s	0.8	221	73/100
Classic C2	Star	'The peripheral men funnelled information to the centre where an answer decision was made and the answer sent out. This organisation had usually evolved by the fourth or fifth trial and remained in use throughout' p. 42.	32s	1.2	215.2	55.5/100
Classic C2	Chain	'information was usually funnelled in from both ends [...] whence the answer was sent out in both directions. The organisation was slower in emerging than the Y's or the [Star's], but consistent once reached' p. 42.	53.2s	1.8	277.4	74/100

From Table 3.1 it emerges that whilst the classic C2 networks (the Star, Chain and 'Y') show good performance in terms of metrics like error and task completion time, as Chapter 2 hinted at, these are no longer the only criteria for judging organisational success. It could be argued that some of the attributes shown by the active, leaderless and seemingly more disorganised Circle archetype may be equally desirable in achieving a better balance between conflicting and dynamic requirements. Leavitt's study is not ideally designed to show this as it employs what can be regarded as a very deterministic kind of task, one that would tend to favour the classic C2 networks. This gives us cause to wonder what the results would be if the desired outcomes were not error or task completion time but innovation, adaptability and emergence. In other words, how would a circle or NEC-like organisation perform on an altogether less deterministic task? It is to this that the discussion now turns.

Experimental Issues

The purpose of this chapter is to take the anecdotal evidence observed in the field and try to recreate, if not the exact situation, then at least an experimental analogue in the laboratory using a paradigm very similar to Leavitt's study. The advantage of lab-based research, of course, is the degree of control that can be imposed, control that was almost entirely lacking in the field study alluded to at the beginning of the chapter. However, caution needs to be exercised. Paradoxically, too much control could conceivably prevent the emergence of the adaptive behaviour for which explanations are sought. Experimental control conflicts with ecological validity which suggests that for these types of problems perhaps a different approach to experimental design needs to be adopted.

In the present study what might be referred to as a classic hierarchical C2 organisation was created within a simulated environment, then pitted against a peer-to-peer NEC counterpart, both of which contained live actors (as per Leavitt's study) who had to operate within a complex, dynamic and adaptive scenario (unlike Leavitt's study). Both conditions thus represent frameworks that people undertake a common task within, but different constraints apply to the different conditions. For example, there is relatively little in the way of rigid specification of experimental procedures in the NEC condition (the focus is on outcomes or 'effects' rather than the actions required to bring them about). In the NEC condition the technological infrastructure is also configured to facilitate peer-to-peer interaction, should the human role incumbents choose to make use of it. The opposite is true for the classic C2 condition. Here there is a high degree of 'scripting of tasks' and a more constrained technological infrastructure within which this occurs.

The research question, therefore, shifts slightly. It focuses on the performance of the incumbents in the different organisations and how that performance changes as they adapt to their context. In a break from traditional human-centred approaches whereby the interaction (and subsequent representations) is generally static (Lee, 2001; Woods and Dekker, 2000) we, like Leavitt to some extent, assume it to be dynamic. Apart from Leavitt's study there is a good basis for taking this position. Patrick, James and Ahmed (2006), for example, recognise the 'unfolding' nature of command and control. They state that, 'A critical feature of command and control in safety critical systems is not only the dynamically evolving situation or state of the plant but also the fluctuating responsibilities, goals and interaction of team members' (p. 1396). Our experimental design needs to take this inviolable aspect of command and control into account.

Hypotheses

Combining Leavitt's earlier findings with the rubric of sociotechnical systems theory, along with what has been observed of live-NEC so far, the following exploratory hypotheses are put forward:

- Hypothesis #1 – Differences among patterns: Leavitt's findings lead us to expect that the C2 network will favour the commander and will develop in such a way as to maintain this pattern. Sociotechnical systems theory leads us to expect from the NEC network a more adaptive network that does not favour to the same extent one individual role incumbent.
- Hypothesis #2 – Time: Leavitt's paper leads us to anticipate the classic C2 network starting off performing the task slower than the NEC network, but catching it up and overtaking it.
- Hypothesis #3 – Messages: Leavitt's paper leads us to expect that the intensity of communications in the NEC condition will be higher than that for the classic C2 condition.
- Hypothesis #4 – Errors: Overall, Leavitt's findings lead us to anticipate better error performance from the NEC condition as compared to the classic C2 condition.
- Hypothesis #5 – Questionnaire Results: Leavitt's findings, combined with the insights provided by sociotechnical systems theory, lead us to anticipate the NEC condition leading to a better subjective experience for those at work within it as compared to the classic C2 condition.
- Hypothesis #6 – Message Analysis: Leavitt's findings lead us to expect a qualitatively different type of communication to occur between NEC and classic C2 conditions. Specifically, that the interactions will be informationally richer for the NEC condition.

The insights provided by sociotechnical systems theory, combined with the observations of live-NEC and the experimental results from Leavitt (1951), lead us to anticipate better initial conditions for more effective adaptation under NEC conditions. The 'model', however, needs to be run in order to find out.

Method

Design

The experimental task is based around a simplified 'Military Operations in Urban Terrain' (MOUT) game called 'Safe houses'. The game creates a dual-task paradigm. The first task involves a commander managing two live 'fire teams' as they negotiate an urban environment en route to a 'safehouse' (note that the experiment took place in a civilian area; fire teams were not actually armed). The second task involves the commander managing the activities of ten further simulated fire teams within a much wider area of operations. The two tasks interact such that success in one does not necessarily connote success overall. It falls to the commander to effectively balance task demands under the independent, between subjects variable of command and control 'type', which has two levels: NEC and C2.

The study is longitudinal in nature. The two teams (NEC and C2) separately undertook a total of 30 iterations through the same dynamic task paradigm and a simple form of time series analysis was employed to reveal the underlying, high-level 'adaptive model' embedded in the data. Participant matching, training and randomisation were employed to control for individual differences, order effects and task artefacts.

Several 'socio' aspects of performance were measured. Thirty separate social-network analyses were undertaken, with the network metrics so derived being subject to the same form of simple regression-based time series analysis in order to show how the pattern of communication changed over time. This combined with an analysis of communications quantity and content (also dependent variables), the latter being facilitated with a pre-defined communications taxonomy. The final 'socio' dependent variable was the subjective experience of the role incumbents, which was measured over seven time intervals using a modified self-report cohesion questionnaire.

Several 'technical' aspects of performance were also measured. Good performance in these terms equates to the shortest task completion time, all en route Target Areas of Interest (TAI's) correctly located and effected, and a high ratio of enemy to friendly agents neutralised.

Participants

There are five principle roles in the study, three of which were occupied by experimental participants (all aged 21). The remaining two roles were filled by the experimenters. The experimental roles are shown in Table 3.2.

Materials

Command and Control Microworld   Figure 3.2 presents a visual representation of the command and control microworld within which the safehouses game was implemented.

Table 3.2   Experimental participants and their roles

Role	Role purpose
Commander	The incumbent of this role was in charge of both fire teams within the primary task, providing guidance and strategy as required. They were also responsible for the larger strategic secondary task.
Fire Team Alpha	This participant was located within the live battlespace and communicated to the commander, and depending on experimental condition, the other fire team as well, by using an XDA mobile device and MSN Messenger™. The XDA device also enabled the fire team to be live-tracked and represented on the commander's command-wall representation of the battlespace.
Fire Team Bravo	This participant had the same role and capabilities as Fire Team Alpha.
NEC System Operator (Performed by member of experimental team)	The NEC System Operator dealt with the experimental aspects of the commander's primary task (i.e., managing the fire teams) as well as the NEC system itself. The NEC System Operator effectively 'drove' the NEC command-wall system, receiving requests to add/append data to the live maps from the commander and helping them to use the system themselves. The system operator also supplied strategic injects according to pre-set rules. In the NEC condition, the experimenter also provided situational updates to all team members (ensuring that 'everyone' knew 'everything').
Enemy (Performed by member of experimental team)	This individual, like the NEC system operator, was another member of the experimental team but this time physically isolated from the battlespace, the commander and the NEC system operator. They were in charge of playing the commander in the secondary task, thus they controlled enemy actions in a 'Wizard of Oz' fashion.

The system operator and commander were co-located in the laboratory. Both sat with a clear view of the so-called 'command wall' which was a large projection screen displaying a modified Google Earth™ representation of the battlespace. This visualisation was annotated with the live position of the fire teams (via GPS) as represented by an appropriate icon. This visualisation was supplemented by a separate large screen displaying a 'planning window' which contained a simple map-based representation of the battlespace environment with a grid-square coordinate system superimposed over the top of it. The planning window allowed the system operator and commander to add, delete

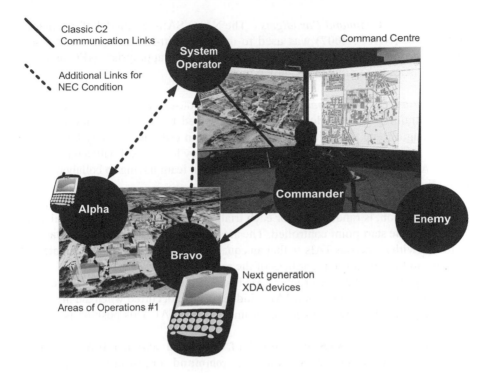

Figure 3.2   Command and control microworld

and move objects as required by the primary and secondary tasks. Any such changes were instantaneously represented on the main visualisation display. The planning window was populated by the experimenter (before the respective experimental condition commenced) with all the required TAI's and enemy icons positioned according to a preset template which was randomly selected for that trial. The commander and system operator had separate work stations and their own computer, and communicated purely through text based means (using MSN Messenger™).

Both fire teams (Alpha and Bravo) were located in the physical battlespace, away from the co-located commander and system operator. The fire teams carried an XDA mobile-phone device and it is this that permitted them to be live-tracked using GPS. The XDA device also allowed each fire team to communicate with each other as well as, although depending on experimental condition, the commander in the control centre. Again, all such interactions were performed textually via MSN Messenger™. Fire teams could also use the XDA to add icons into their own miniature version of the digital map which would then simultaneously appear on each others screens as well as the main visualisation window in the command centre. The miniature digital map shown on the fire teams' XDA screens was also used for navigation purposes.

Command and Command Paradigms   The NATO SAS-050 model of command and control (NATO, 2007) was used to design Leavitt-like networks and a set of organisational constraints that combined to enable appropriate NEC and C2 characteristics to be created, as shown in Table 3.3.

'Safe houses' Game (Primary Task)   The safe-houses game was inspired by a military training guide (MoD, 2005) and involved the fire team choosing and negotiating a route through an urban battlespace in order to correctly locate and effect a so-called safe house. In addition to this they had to deal with a number of Target Areas of Interest (TAIs) en route, with each team having to collaborate by providing cover for the other.

Both fire teams start from the same location. The location of the safe house, the final destination, is randomised for each trial but with distance (and potential task time) from the start point controlled. The area of operations for the primary task is scattered with numerous TAIs so that an equal number of them will be encountered en route to the safehouse regardless of the route chosen to get there.

Firstly, each of these 'en route' TAIs had to be correctly located by one of the fire teams. Correct location of a TAI is judged to have occurred when the fire team takes up position at the same grid coordinates as the TAI. This position is shown

Table 3.3   **NATO SAS-050 Model of Command and Control was used to design NEC and C2 command organisations with the appropriate characteristics**

	NEC	C2
DISTRIBUTION OF INFORMATION	BROAD: fire teams are provided with regular situation updates from the system operator in addition to being able to interact directly with their counterparts. 'Everyone' knows 'everything'.	TIGHT: the commander is the only individual with an overall view of the situation. The fire teams' had a local view of their immediate location but in all other respects worked in isolation. 'Everyone does not know everything'.
PATTERNS OF INTERACTION	DISTRIBUTED: all team member roles could speak to each other independently (there was no communications hierarchy).	HIERARCHIAL: the fire teams could speak to the commander but not directly to each other.
ALLOCATION OF DECISION RIGHTS	PEER TO PEER: collaborative working encouraged and facilitated by effects based instructions and peer-to-peer communications infrastructure.	UNITARY: autonomy, authority and discretion rested with the commander only.

on the XDA device's digital map display. Missing out a TAI by failure to locate it will result in the offending fire team being removed from the mission and having to return to the start point for the remainder of the trial.

Secondly, assuming the en route TAI has been correctly located by the fire team, who will now be standing on the requisite grid square coordinates, the fire team then has to 'effect' it in order to make it safe for the other fire team to continue on their route. Although the location of the TAI is known by the fire team and commander a priori, what is not known is what form it actually takes and the most appropriate way to 'effect' it. This can only be judged by the fire team who are on the ground and are able to make that assessment based on a number of simple local characteristics. These are as follows:

- If the TAI coordinates correspond to the side of a building over three storeys high then a 'yellow effect' will neutralise it which is signified by the relevant fire team using their XDA to place a yellow icon on the appropriate grid coordinate.
- If the TAI is located on the side of a building less than three storeys high, then a 'blue effect' will neutralise it. This is signified by a blue icon being placed.
- If the TAI is located in the middle of a thoroughfare then a 'pink effect' will neutralise it, signified by a pink icon.

After confirmation that this information has been received, the relevant fire team will hold in this position, providing cover for the other fire team as they make their way to the next TAI. This 'leapfrogging' effect continues until, finally, the safehouse itself is located and affected in exactly the same way. It should be pointed out that in the NEC condition this leap-frogging is facilitated by the fire teams' being able to communicate directly with each other via their XDAs; in the C2 condition, however, communication (and instructions) have to pass through and/or come from the commander.

In order to further encourage the need for communication and interaction there is also a degree of built-in ambiguity in the positional data. This means that part of the adaptive process of the entire team is to figure out 'work-arounds' and modes of operation that enable these ambiguities to be resolved in whatever way is found to be most efficient.

The need for good time and accuracy performance is embedded in the scenario by two simple game-play expedients. As mentioned before, if the wrong location is chosen or the TAI is ignored then the fire team allocated to it fails the mission and has to sit out the remainder of the trial. If the right location but wrong effect is applied then the fire team's attrition score, which starts at five and denotes 'full strength', is decremented. If the attrition score for a particular fire team reaches zero, their effectiveness is neutralised and they cannot continue the task. The attrition score is not just affected by accuracy but also speed and time. Five time-activated attrition-injects occur throughout the fifteen minutes allotted to the trial;

these cause both fire team's attrition scores to be decremented the longer they spend in the battlespace.

What appears to be a relatively complex set of rules becomes considerably simplified as far as the experimental participants are concerned. The system operator (who is a member of the experimental team) undertakes all game-play management tasks such as maintaining the formal record of location accuracy, whether the right effect has been applied, the attrition score, enactment of the time-based attrition-injects, and communicates all of this to the commander as required.

Secondary Task (Commander)   Whilst the first task concerns the activities of a live-fire team in a live environment, the secondary task of the commander is simulated and occurs in a much wider area of operations (AO). The boundaries of this are six miles in either direction from the AO of the first task. This larger strategic mission relies on the commander playing a competitive game against a simulated enemy played by the second experimenter, who is acting in a wizard of oz fashion. Thus 'live' (small AO) and 'simulated' (large AO) elements are played simultaneously on the same NEC system.

So, whilst the live fire team progress towards their primary target, within their smaller AO, wider enemy activity is taking place all around which, if permitted to continue, will eventually impinge on the primary task. Such impingement, under certain specified conditions, will mean that the primary task fails regardless of the efficacy of the live-fire teams. The onus is thus on the commander to manage both tasks effectively.

In the commander's wider area of operations there are fifteen enemy elements/ icons dispersed randomly around the environment. These icons are placed into the main visualisation window of the NEC system according to a random script for that trial, by the experimenter acting in the 'enemy' role. Enemy icons, due to the asymmetry of most MOUT-type situations, are free to move across the battlespace at will with no area being restricted to them. On the other side, the commander has 25 friendly-force icons/elements all massed in a defined 'green zone'; this is their starting position for the game. Although far more numerous, the friendly icons are not permitted to enter predefined 'sensitive areas'.

Apart from these constraints, the game plays like a virtual game of draughts in which the grid square coordinate system of the map serves as a form of draughts board. Only one icon can be moved, one square at a time (in any direction) in enemy reaction, friendly counter action, enemy counter re-action, and so on in sequence. If the friendly icon enters a grid square occupied by an enemy icon then the friendly icon wins. If the enemy icon enters a grid square occupied by a friendly icon, the enemy wins. The commander controls his friendly pieces through textual means (again with MSN Messenger™) by telling the experimenter acting in the enemy role which coordinates he would like his particular friendly piece to move to. The experimenter playing the role of enemy updates the icons/

map accordingly and keeps a record of the secondary task score, which is the ratio of enemy/friendly pieces removed from the game.

Cohesion Questionnaire   A modified version of the Combat Platoon Cohesion Questionnaire (Siebold and Kelly, 1988) was deployed in order to provide some measure of the role incumbents' subjective experience of the two structures (NEC and C2). The questionnaire was shortened to fifteen items, all of which required a response along a 7-point Likert scale. A score of one equates to the maximum score on any given cohesion question, 7 the minimum. The individual question items, and the corresponding mapping to Hackman and Oldman's (1980) core job-characteristics and the sociotechnical concept of 'meaningfulness of tasks', are shown below in Table 3.4.

Table 3.4   Question items drawn from the Platoon Cohesion Questionnaire (Siebold and Kelly, 1988) mapped onto Hackman and Oldman's (1980) core job characteristics to create a much simplified assessment method

Question item	Core job characteristic				
	Skill	**Task identity**	**Task significance**	**Autonomy**	**Feedback**
These items rated from 'not at all important' to 'extremely important'					
Loyalty to the team		■			
Taking responsibility for their actions and decisions				■	
Accomplishing all assigned tasks to the best of their ability			■		
Commitment to working as members of a team		■			
Dedication to learning their job and doing it well	■				
Personal drive to succeed in the tasks			■		
Taking responsibility to ensure the job gets done				■	
The team can trust one another		■			
Feel close to other team members		■			

Table 3.4   *Concluded*

Question item	Core job characteristic				
	Skill	**Task identity**	**Task significance**	**Autonomy**	**Feedback**
These items rated from 'very much/well/always etc' to 'very little/poorly/not at all etc'					
How well do the members in your team work together?		▓			
To what extent do team members help each other to get the job done?				▓	
To what extent do team members encourage each other to succeed?			▓		
Do the members of your team work hard to get things done?				▓	
The chain of command works well?					▓
Everyone is well informed about what is going on?					▓

Procedure

Training Phase (Day #1)

The aims and objectives of the experimental task were introduced to the external participants along with health and safety preliminaries and informed consent. Detailed instructions on the task were then provided to all participants, supplemented with demonstrations and hands-on examples. The experimenter then used the pre-populated command-wall to begin the first full trial which was identical in all respects to the experimental trials but serves initially as a practice (both teams are measured subsequently as an internal check on concordance between them).

Experimental Phase (Day #2–8)

With the participants familiar with the task paradigm, the teams begin to undertake the repeated iterations of it. Issues and questions are dealt with before the trial starts and during it if required. The task gets underway and the participants interact in the manner prescribed by the organisational type they are currently working within. They have a maximum time of fifteen minutes within which to complete the task.

Results

To briefly recap, two teams took part in a simulated MOUT mission over 30 successive iterations. The analysis focuses on how the different constraints of NEC and C2 influenced the direction of team adaptation and performance over time. In broad terms it is hypothesised that NEC will provides better conditions for adaptation to the complex dynamic experimental task. A simple form of time series analysis will be deployed in order to uncover the underlying theory behind the data and thus put this supposition to the test. The results that follow are compared to Leavitt's original findings and presented under that paper's original sub-headings:

Differences Among Patterns

Of the network structures more pertinent to classic C2 Leavitt (1951) writes: 'The star operated in the same way […] The peripheral men funnelled information to the centre where an answer decision was made and the answer sent out. […] The Y operated so as to give the most-central position […] complete decision-making authority. […] In the chain information was usually funnelled in from both ends […] whence the answer was sent out in both directions' (p. 42). Of the network structure most pertinent to NEC the following is noted: 'The circle showed no consistent operational organisation' (Leavitt, 1951, p. 42). Our findings show the following.

The structure of communications (as in who can speak to whom) is less constrained in the NEC condition than it is in the C2 condition. In the NEC condition everyone can speak to anyone (if they want to), whereas in the C2 condition the structure of communications is fixed and all communications have to pass through the commander. Social Network Analysis (SNA) lends itself well to examining structural determinates of communication like these. Briefly, a social network can be plotted by examining 'who' is communicating to 'whom'. The presence of a communication establishes a link between any actors that are communicating with each other, creating its own structure in the form of a social network. The network can then be examined mathematically to identify key characteristics such as sociometric status (who, based on their degree of interconnection, can be regarded as a key agent) and density (how densely interconnected the network as a whole is). In the present case, density and sociometric status did not behave in a way that made the linear regression method of time series analysis tenable. In terms of both these measures the regression diagnostics showed only weak (and non-significant) associations between them and trial interval. Likewise, the regression ANOVAs failed to detect a meaningful and/or significant effect in terms of the data behaving linearly. This form of modelling is therefore substituted for a more conventional cross-sectional analysis.

The simplicity and constraints imposed on both the NEC and C2 social networks are clearly evident in the behaviour of the density metric. In both cases (albeit during different trial intervals) the density figure only deviated from a fixed point on three occasions. This behaviour makes non-parametric tests appropriate. The results show that the mean density figure in the NEC condition (M = 0.97, SD = 0.1) was somewhat higher than the C2 condition (M = 0.61, SD = 0.19). A Mann-Whitney U test goes on to suggest that the NEC social network is significantly more dense (U = 40.5; exact p < 0.01) than the C2 network, meaning that the opportunity for peer-to-peer collaboration was taken by the study participants who, remember, were under no obligation to use it. The approximate effect size of this finding is r_{bis} = 0.25, a relatively small effect yet one consistent with the small size and simple characteristics of the networks under analysis. The findings for sociometric status are presented in Table 3.5.

Table 3.5   Results of analysing sociometric status in relation to NEC and C2 conditions

Agent	Type	Mean sociometric status	Standard deviation	U statistic	Exact p	Approx effect size (r_{bis})
Commander	NEC	7.88	3.27	52.5	< 0.0001	0.26
	C2	20.95	8.99			
Alpha	NEC	11.92	2.73	417	= 0.32	0.04
	C2	12.02	5.39			
Bravo	NEC	12.13	3.29	222.5	< 0.001	0.22
	C2	8.93	4.12			

The pattern of results shown in Table 3.5 paints quite a coherent picture. As anticipated, in the C2 condition the commander has significantly elevated status compared to the NEC condition (mimicking the results observed for Leavitt's Star, Y and Chain networks). The NEC condition, on the other hand, was not as inconsistent as Leavitt's study might have suggested. Faced with this particular complex, dynamic and adaptive task the opportunity for peer-to-peer interaction was seized upon and exploited. This, in combination with elevated levels of communication traffic, led to higher network density and a more even spread of sociometric status among the participating agents. In sociotechnical terms this data is suggestive of a greater degree of internal leadership and autonomy, with all agents able (as in fact was the case) to communicate and participate with no single agent becoming particularly dominant.

Task Performance Time

In regard to task performance time Leavitt (1951) writes: 'A measure of the fastest single trial of each group indicates that the star [i.e., classic C2] was considerably faster (at its fastest) than the circle [i.e., NEC]' (p. 43). The adaption curves presented by Leavitt also trace a distinct pattern, one in which the circle/NEC-like archetype starts off performing the task more quickly but the curve for the star/classic C2-like archetype eventually overtakes it. In other words, the curves cross with the result that the star/classic C2 archetype becomes quicker than the circle/NEC archetype even though it did not start out that way. Our simple time series findings are more or less identical in this regard.

Both teams (C2 and NEC) were measured in terms of how long it took the live-fire team to perform their task. When this first task was complete then both tasks of the dual-task paradigm were halted. The maximum amount of time that was allowed to be spent on the task was 15 minutes (900 seconds). As one would expect, over the course of the 30 iterations both teams speeded up considerably and continued to do so for every trial. This we can consider as a linear first order effect which we can analyse using linear regression as a method of time series

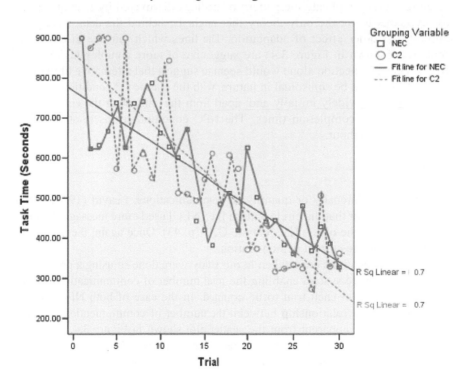

Figure 3.3   Scatter plot showing the regression lines for both NEC and C2 conditions in relation to task time

analysis. A strong association between task time and trial was obtained for both conditions (NEC $r = -0.84$ and C2 $r = -0.85$), both of which were significant at beyond the 1 per cent level. Furthermore, the regression ANOVA supports the hypotheses that this association is linear in nature for the NEC condition: $F(1,28) = 64.74$; $p < 0.0005$ and $F(1,28) = 73.53$; $p < 0.0005$ for the C2 condition.

The linear regression model fitted to the data accounted for 70 per cent of the variance in the NEC condition (Adjusted $R^2 = 0.70$) and 72 per cent of the variance in the C2 condition (Adjusted $R^2 = 0.72$). Both values represent a large effect size and both regression models were statistically significant to beyond the 1 per cent level. The regression equation, however, differed between the two conditions. The intercept for the C2 condition was at $b_0 = 862$ seconds, somewhat nearer the maximum value of 900 seconds permissible for the task than the NEC condition, whose intercept was at $b_0 = 762$ seconds. However, the regression line for the C2 condition had a slightly more precipitous slope than that for the NEC condition, $b_1 = -14.26$ compared to $b_1 = -19.73$; thus, despite the higher intercept, the regression lines actually crossed at trial 17 meaning that by trial 30 the regression model predicts the task being completed in 270 seconds for the C2 condition compared to 334 seconds for the NEC condition (approximately a minute faster). These results, therefore, exactly replicate the pattern of findings discovered by Leavitt (1951). However, this is not the only theory that might lie behind the data, merely the simplest first order effect of adaptation. The lines which join the discrete data points (also shown in Figure 3.4) are suggestive of more complex higher order effects. Visual inspection alone would seem to suggest that alternative fit lines for C2 and NEC could be sinusoidal in nature, with the classic C2 condition seeming to oscillate quite widely initially and, apart from the spike near the end, homing in on faster task completion times. The NEC condition seems to show wider oscillations throughout.

Messages

In regard to the intensity or quantity of communications, Leavitt (1951) writes that, 'It seems clear that the circle pattern [i.e., NEC] used more messages to solve the problem than the others [i.e., classic C2].' (p. 43). Once again, these findings are replicated in the current demonstration.

All communications undertaken in our study were done so using a proprietary messenger-software system enabling the total number of communications events that occurred within each trial to be counted. In the case of both NEC and C2 conditions a linear relationship between the number of communications and the number of trials is apparent from the scatter plot shown in Figure 3.4, at least to the extent that values can be seen to generally decrease with time. This is the type of linear first order effect seen above for task time. A relatively strong negative association between the number of communication exchanges and trial interval is in evidence: $r = -0.43$ (NEC) and -0.72 (C2), both of which are statistically significant beyond the 1 per cent level. This reflects the general trend towards

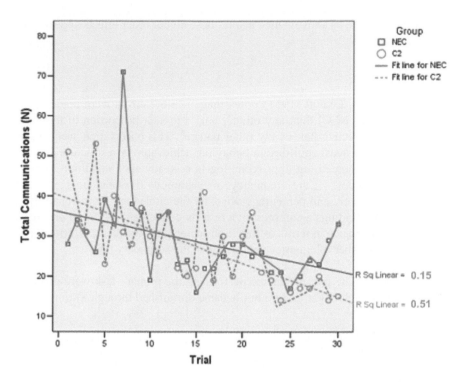

Figure 3.4   Scatter plot showing the regression lines for both NEC and C2 conditions in relation to total communications

fewer communications events as the number of trials increases, with this trend being more pronounced for the C2 condition. According to the regression ANOVA there is good statistical support for the linearity observed in the data: $F(1,28) = 6.18/29.56$; $p<0.05$ (NEC) and $p<0.0005$ (C2).

The regression models fitted to this data differ in respect to the amount of variance they explain. The variance explained in the NEC condition is of the order of 15 per cent (Adjusted $R^2 = 0.15$) whereas the model constructed for the C2 condition explains substantially more at 50 per cent (Adjusted $R^2 = 0.50$). In other words, the number of communications in the C2 condition behaves as though it is attracted more linearly towards lower values than the NEC condition. Indeed, Figure 3.4 shows that the intercepts (b_0) cross. The C2 condition starts off with more comms events than the NEC condition (40.2 comms events versus 36.08) but the higher regression coefficient ($b_1 = 0.51$) leads this situation to invert with time. In formally rational terms (see Chapter 2) C2 could certainly be viewed as 'efficient', but fewer communications may not be better. Note that both regression models (C2 and NEC) are demonstrative of a large effect size and are both statistically significant beyond the 5 per cent level. Reference to the lines traced by joining the data points shows the existence, potentially, of more complex higher order effects, as above. Once again, those higher order effects appear to be sinusoidal in nature.

Task Errors

In regard to errors, Leavitt (1951) notes that: '...more errors were made in the circle pattern [i.e., NEC] than any other', but, 'a greater proportion of them (61 percent) were corrected than in any other pattern'. This observation seems to be consistent with the quasi equilibrium behaviour achievable by a structure that can maintain itself in something approximating to a steady state condition, despite external disturbances. '...the frequency of unanimous five-man final errors is lower, both absolutely and percentage-wise, for the circle than for the chain [i.e., C2]' (p. 44). Our findings again present a broadly concordant pattern, more so for the teams' performance in terms of 'attrition scores' than for their performance in relation to the number of enemies neutralised in the secondary task.

Attrition scores   The live-fire teams performing the primary task were given an attrition score which began at five but became diminished through a) time-based

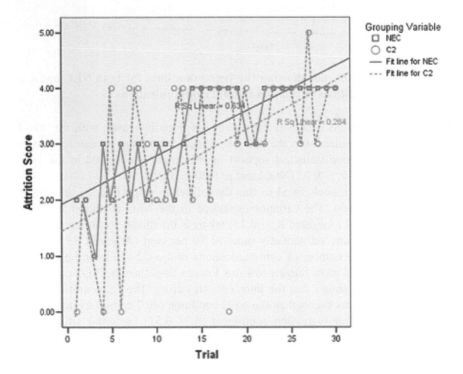

Figure 3.5   Scatter plot showing the regression lines for both NEC and C2 conditions in relation to attrition score

injects, so the longer that was spent on the task the more chance there was of having the score decremented, and b), if locations of TAIs and markers placed at them were inappropriate and/or inaccurate. Simply put, a high attrition score connotes better performance.

In both conditions the attrition score is positively correlated with the number of trials undertaken, $r = 0.8$, $p < 0.01$ for the NEC condition and $r = 0.53$, $p <0.01$ for the C2 condition, meaning that generally speaking error performance improved over time as one would expect. The regression ANOVA supports the hypothesis of linearity in both cases: $F(1,28) = 48.55$; $p<0.01$ for NEC and $F(1,28) = 11.09$; $p<0.01$ for C2. Note that despite the statistical significance of the regression diagnostics the C2 condition possess less statistical power in terms of its associative performance (r), linearity (F) and also in the amount of variance explained by the regression model ($R^2 = 0.26$ compared to NEC's 0.62).

In model terms the regression coefficient (the slope of the regression line) was similar for NEC and C2, being 0.08 and 0.09 respectively. The intercept values were, however, different. The C2 model starts with worse error performance with a lower attrition score of 1.5 and maintains a subordinate position to NEC (whose intercept is 1.98) for the duration of the study. Both of these regression models were statistically significant to beyond the 1 per cent level. Despite C2's favourable evolution towards faster task completion times it seems that it is not all good news. Accuracy suffers. This finding accords with Leavitt's.

Enemies neutralized This factor relates most strongly to the commander's performance in the secondary task (within the wider area of operations). It can be immediately noted that the commander was able to manage these competing tasks quite satisfactorily, with the secondary task at no time causing the premature cessation of the primary task.

The assumptions underlying a linear approach to time series analysis are not met in the case of friendly-versus-enemy capture ratio. In other words, trial number or task iteration appeared not to be a good predictor of this factor's performance. In both cases only small ($r = 0.3/0.24$) correlations were detected for NEC and C2 respectively, albeit statistically significant. However, the fact that the resultant regression model only explained around 6 to 9 per cent of the variance in the data ($R^2 = 0.09/0.06$), the hypothesis regarding linearity was not supported: $F(1,28) = 2.75/1.64$; $p = ns$, furthermore, that both regression models failed to reach significance ($p = ns$) means that this form of analysis can be abandoned. Given the lack of a linear relationship between trial and capture ratio a simple cross-sectional approach can be taken.

Even here, an independent samples t-test failed to detect a statistically significant difference in enemy/friendly capture ratio between NEC and C2 ($t = 1.48$; $df = 58$; $p = ns$). Given that such a test possesses in excess of 80 per cent power to detect medium effect sizes or larger, and that only a very small effect was actually detected ($r_{bis} = 0.04$), means that there is a good degree of confidence in stating that the constraints imposed by both NEC and C2 conditions are not a particularly meaningful determinant of enemy-versus-friendly capture ratio. In other words, C2 evolves towards faster task completion times, but with poorer accuracy, but both NEC and C2 are comparable in terms of the numbers of friendlies and enemyies captured. In this regard, as a metric for errors this particular manipulation/measure did not mirror Leavitt's findings.

Questionnaire Results

In regard now to the ostensibly 'socio' aspect of system performance, the feelings of those at work within different structures, Leavitt (1951) notes: '…we find the order circle, chain, y, star, with circle [i.e., NEC] members enjoying their jobs significantly more than the star members [i.e., classic C2]' (p. 44). Once again our results paint an identical picture.

Cohesion is a subjective experience of team working and an emergent property of the command organisations under analysis; like the other factors, it is a product of the constraints of the organisation within which the team were working. The study was spread out over 7 days and the 15 item cohesion questionnaire was administered at the end of each day to every team member (this was the shortened version of the Combat Platoon Cohesion Questionnaire mentioned earlier; Siebold and Kelly, 1988). The results were summed to provide a quick and simple measure of changes in subjective experience. Linear regression was used to diagnose the underlying model in the data. The results of applying this technique to the NEC cohesion data are promising. All the model diagnostics proved favourable. In the NEC condition team cohesion was strongly and significantly associated with trial interval ($r = -0.77$; $p < 0.05$), assumptions of linearity were supported ($F(1,5) = 7.16$, $p<0.05$) and the regression model explained a substantial amount of the variance in the data ($R^2 = 0.59$). As Figure 3.6 shows, the intercept (b_0) occurred at a cohesion score of 61.43 (out of a maximum of 105) with the slope of the regression line (b_1) representing -1.86. Note that lower scores represent an improvement in subjective experience. As a result, the model suggests that the structural determinates of the NEC condition are associated with meaningful, linear improvements in the self-rated experience of those working within its constraints. The results for the C2 condition, on the other hand, are less promising.

In the C2 condition the regression diagnostics suggest that there is little in the way of association between team cohesion and trial interval ($r = -0.08$; $p =$ ns), assumptions of linearity are not supported ($F(1,5) = 0.03$; $p =$ ns) and the model explains next to nothing of the variance in the data (Adjusted $R^2 = -0.19$). Whilst team cohesion can be more than adequately modelled in linear terms for the NEC condition, with that model predicting meaningful improvements in team cohesion over time, the data for the C2 condition is quite different. A similar meaningful, linear relationship between team cohesion and trial interval does not exist, as per Leavitt's original findings.

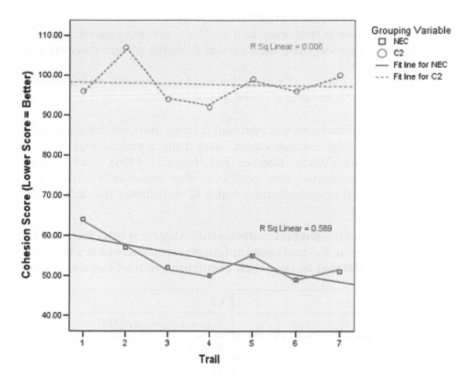

Figure 3.6   Scatter plot showing the regression lines for both NEC and C2 conditions in relation to team cohesion

Message Analysis

Leavitt (1951) notes that not only does the quantity of communications change depending on 'communication pattern' but so to does the type of communication. In particular, 'Circle members send many more informational messages than members of the other patterns [...] Circle members also send more answers' (Leavitt, 1951, p. 44). By now it is perhaps not surprising that we, too, find differences in communication type contingent on organisational structure and in a manner not inconsistent with that just described.

Verbal exchanges between team members were not just recorded, transcribed and counted, but also categorised according to Bowers et al.'s (1998) communications taxonomy as being:

- Factual: 'objective statement involving verbalized readily observable realities of the environment, representing "ground truth".'
- Meta-query: 'request to repeat or confirm previous communication'.
- Response: 'statement conveying more than one bit of information' (i.e., comprising of more than simply 'yes/no').

- Query: 'direct or indirect task-related question'.
- Action: 'statement requiring team member to perform a specific action'.
- Acknowledgement: 'one bit statement following another statement (e.g., "yes", "no")'.
- Judgement: 'sharing of information based on subjective interpretation of the situation' (Cuevas et al., 2006, p. 3–4).

Each individual category was analysed in terms of its relative contribution to the aggregate 'total communication' score using a methodology based on multiple regression (Warm, Dember and Hancock, 1996). Each category is ascribed a standardised beta coefficient that numerically expresses its contribution to total communications which in turn enables the rank ordering shown in Table 3.6.

Table 3.6   Standardised beta coefficients showing the relative factor loading of the individual communications categories within a regression model that assesses their contribution to total communications

NEC	C2
Response (0.36)	Acknowledgement (0.36)
Query (0.32)	Response (0.29)
Factual (0.22)	Action (0.26)
Acknowledgement (0.21)	Query (0.25)
Meta-query (0.17)	Meta-query (0.12)
Action (0.16)	Judgement (0.08)
Judgement (-0.06)	Factual (0.07)

The findings from this analysis show that NEC is not only relatively communications intensive but also that the communications flowing around the network are different in 'type' compared to the C2 condition. Taking the top three contributors, it can be stated that the content of communications for the NEC condition are characterised by exchanges 'conveying more than one bit of information', 'direct or indirect task-related question[s]' and exchanges concerning the 'readily observable realities of the environment'. The C2 condition is characterised by one bit 'yes/no' statements and slightly elaborated versions of the same. It is also characterised by 'statement[s] requiring team member[s] to perform a specific action'. The effect of the hierarchical interactions and 'action based' instructions can, therefore, be clearly seen. Echoing Leavitt (1951) for a further time, the NEC condition is implicated in informationally richer exchanges than C2. This in turn would seem consistent with the adaptive, jointly optimising nature of this network type.

Summary

The results of Leavitt's original paper show that organisational structures containing features consistent with classic hierarchical C2 perform best in deterministic tasks. This study shows that NEC-like structures perform best in complex, dynamic and adaptive tasks. This is clearly evident from the fact that all six exploratory hypotheses find support in the experimental demonstration just described. In Hypothesis #1, the NEC condition *was* more adaptive, albeit ultimately slower than classic C2 (Hypothesis #2). The intensity of communications *was* higher in the NEC condition (Hypothesis #3) *and* informationally richer too (Hypothesis #6). Error performance *was* better for NEC (Hypothesis #4) and in Hypothesis #5 NEC *did* succeed in offering an improved experience for those at work within it.

The interesting experimental issue raised by this demonstration is that the results would have been quite different had the human system interaction been assumed to be stable. As it happens, it is in precisely this instability, combined with the means to create conditions that favour a particular type of 'desirable instability', that the strength of the sociotechnical/NEC paradigm seems to lie. This provides a powerful clue that a further, much more fundamental process is at work here. If the dominant implicit theories that lie behind the organisational design of classic C2 and other rationally designed organisations are wedded to the 'monism' of technology, and sociotechnical systems theory represents the 'dualism' of people and technology, a third factor, subordinate to neither but arising from the *interaction* of both, is represented by 'complexity'. Therefore, in order to move on from practical examples of sociotechnical interventions in the business arena (the first bridgehead described in Chapter 2) and the smaller-scale experimental versions of NEC (the bridgehead described in this chapter) onwards into 'real-life', large-scale, in-service NEC, we need to establish one further conceptual bridgehead: to complexity and the triad of the social, the technical and the complex.

Chapter 4

Complexity and Human Factors

Aims of the Chapter

Since 1958, over 80 journal papers from the mainstream human-factors literature have used either the words 'complex' or 'complexity' in their titles. This chapter addresses the need to define exactly what is meant when these terms are invoked and to argue that this is the fundamental nature of the problem to which NEC and sociotechnical systems theory are a response. This question assumes a surprising degree of more pressing relevance because over 90 per cent of the papers incorporating 'complexity' in their titles have been published since 1990. Clearly, something profound seems to be happening in regard to how human factors problems are changing and the types of tools and techniques needed to cope with them.

The current chapter distils the fundamental concept of complexity through three overlapping themes: 1) the attribute view, which leads to a multi-dimensional problem space through which human factors, as an academic field, appears to be travelling; 2) the complex theoretic view, in which metrics and measures exist to complement established human-factors methods and diagnose at least certain aspects of complexity; and 3), the complex systems research view. This perspective is given particular emphasis because it is in here that we find not just the 'what' of complexity, but also the 'why' and the 'how'.

A number of important caveats are attached to the work in this chapter. Complexity, as a field of scientific enquiry, is itself complex. It has not been possible in this chapter, or indeed this book, to be exhaustive, or to pull through all of the finer points associated with the myriad concepts involved. The chapter necessarily focuses on breadth of coverage rather than depth and it sacrifices strict scientific rigour in favour of innovation. With these caveats in place, the aim of this chapter is to establish an initial model and a conduit for knowledge to flow between the physical sciences (from which the formal study of complexity principally derives) and the human sciences (where the effects of complexity are acutely felt).

Human Factors Complexity

The word 'complex' seems to have first justified a use in the title of a paper published in the mainstream human factors literature as long ago as 1958 (e.g., Chiles, 1958) and since then over 80 journal articles have featured either the word 'complex' or 'complexity' in their titles too. As Figure 4.1 shows, nearly 90 per cent of these have been published since 1990. If complexity is not rising, then

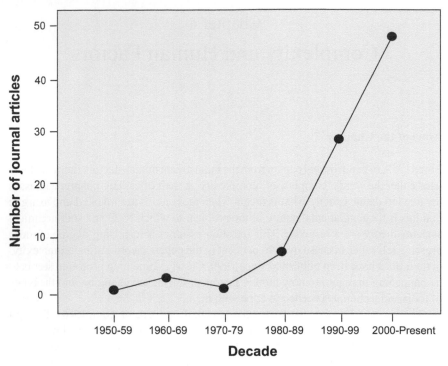

Figure 4.1   **Number of articles in the mainstream, peer-reviewed human factors literature that have either the word 'complex' or 'complexity' in their title (based on a search of *Theoretical Issues in Ergonomics Science, Ergonomics, Applied Ergonomics,* and *International Journal of Cognitive Ergonomics and Human Factors*)**

it is certainly a term that human-factors practitioners find increasingly relevant to the problem domains they now face. The problem is that the wider human-factors literature remains virtually silent on formal definitions, measures, and even awareness of complexity as a distinct and emergent concept in its own right. To date, the precise meaning and implications of complexity have been left for the reader to infer. Judging by Figure 4.1 though, the time has come to try to address this gap in knowledge.

The 'Attribute' View

What do human factors professionals typically mean when they invoke the term complexity? Quite often it is little more than the dictionary definition of the term, that something consists of 'several parts' and is 'involved, intricate or difficult' (Allen, 1984, p. 145; Richardson, Jones and Torrance, 2004). But there is clearly more to complexity than this, and indeed, most articles that use the term seem to be trying to say something much more profound about their particular problem space. Most human factors professionals would probably agree with Woods (1988) that complexity has a number of fairly distinct features, and this we can refer to generically as the 'attribute view' of complexity. The attribute view can be distilled down to five main components.

Multiplicity

The first of these attributes can be labelled 'multiplicity', as in multiple potential causes for extant phenomena and multiple consequences, or 'a great number of interrelated and therefore interacting factors' (Marmaras, Lioukas and Laios, 1992, p. 1224). In the wider human factors literature there are 'multiple people' and a class of paper headed 'complex team tasks' (e.g., Braard, 2001). There are also 'complex multiple tasks' involving 'the control of a large number of interdependent process variables' (Sauer, Hockey and Wastell, 2000, p. 2044) and there are 'complex distributed teams', another popular heading in which these two strands (multiple tasks that are dealt with by multiple people) are brought together more often than not as a result of networked technology like the internet (e.g., Rogalski and Samurcay, 1993). As a dimension of human-factors complexity, the notion of multiplicity also overlaps with that of a 'system' (Naikar, Moylan and Pearce, 2006) and certainly a large number of articles ally themselves to a systems perspective (and use the heading 'complex system' as a result; Gegoriades and Sutcliffe, 2006; O'Hare, 2000; Lo and Helander, 2007; Wei, 2007; Kaber et al., 2001; Hanisch, Kramer and Hulin, 1991; Swain, 1982 etc.). It is important to point out, however, that the term 'complex system', like 'sociotechnical system', has a much more profound meaning than its current use as a descriptive label might at first suggest.

Dynamism

The second attribute is 'dynamism'; that is, to what extent can the system change states without intervention from the user? To what extent can the nature of the problem change over time? To what extent can multiple ongoing tasks have different time spans? (Woods, 1988). The role of time is captured in the numerous articles headed 'complex *dynamic* systems' (e.g., Elg, 2005; Howie and Vicente, 1998; Vicente, et al., 2004; Reinartz, 1993; Canas et al., 2005) in which simulations, micro-worlds and/or other longitudinal methods of the sort similar to that presented in Chapter 3 are common to them all. This could be viewed as a tacit rejection of the non-dynamic representation of the human-system interaction embodied by the more ubiquitous psychology-like cross-sectional study (Lee, 2001). It is certainly recognition of the entirely pragmatic fact of human-factors life that problems rarely possess the clinical levels of control found in the lab, that there is often no 'one right way' to perform a task (Howie and Vicente, 1998) that people will interpret their environment and that there are more experimental hypotheses available than time to conduct the requisite studies. Human-factors complexity, therefore, as well as being an attribute of systems, is also associated with degrees of freedom and change over time.

Uncertainty

The third attribute is 'uncertainty'. Tacit in the literature is the idea that parts, interconnections and dynamism make it difficult to discern final states from initial conditions. Furthermore, these initial conditions seem sensitive to uncertainty in the form of inaccuracy, randomness, and vagueness (Hancock, Masalonis and Parasuraman, 2000). Of particular note is a growing class of paper that overtly recognises this attribute and terms it 'fuzziness'. As a result, there are numerous fuzzy models (e.g., Luczak and Ge, 1989; Karwowski and Ayoub, 1984), ranging from neuro-fuzzy models (Lee et al., 2003) to fuzzy linguistic models (McCauley-Bell and Crumpton, 1997), even a rebuke to the binary, first order logic of category membership; fuzzy signal detection theory (Masalonis and Parasuraman, 2003). If a trend can be discerned from all this, then it is a recognition that artefacts of human factors problems do not always occupy neat orthogonal categories and that it is not always possible to have complete knowledge of a problem or phenomena.

Difficulty

The fourth attribute of complexity is 'difficulty', with the human-factors world reflecting back a prominent component of the dictionary definition. An ergonomic axiom is that complex tasks take longer to learn and are often more demanding to perform (Sauer et al., 2006) with around a third of articles that feature complexity in their title doing so in relation to what they term 'task complexity'. Although complexity can be expressed objectively as a form of 'intrinsic complexity', different levels of which form independent variables in studies (e.g., Sauer et al., 2006), it is much more common within the literature for complexity to reside in the eye of the beholder in the form of 'supposed complexity' (Leplat, 1988) or 'complexity seen from the perspective of the person describing the task' (Cronshaw and Alfieri, 2003, p. 1108; Bar-Yam, 2004b). Interestingly, several articles use the NASA TLX, and by implication workload, as a surrogate for this (e.g., Gregoriades and Sutcliffe, 2006; Braarud, 2001). The higher the workload the more difficult and more supposedly complex the task is held to be. O'Brian and O'Hare (2007) point out that a major source of this difficulty is modern systems that 'increasingly challenge the operator's perceptual and cognitive, rather than physical, abilities' (O'Brian and O'Hare, 2007, p. 1064) attached to which is an unfortunate trend that addresses shortfalls in system-capability with more, not less complexity (e.g., Hollnagel and Wood's 'self reinforcing complexity cycle'; 2005). This, too, is a prominent facet of the literature gathered under the umbrella of 'complex decision making tasks' (Quesada, Kintsch and Gomez, 2005; Canas et al., 2005; Reinartz, 1993; Marmaras, Lioukas and Laios, 1992; Coury and Drury, 1986 etc.).

Importance

The fifth and final attribute of ergonomic complexity is 'importance'. Woods (1991) equates complexity with what is at stake, so for an entity or artefact to be complex in a human factors sense it has to be meaningful. With reference to Figure 4.1 and the apparent growth of complexity, it seems rather improbable to suggest that the domains of anaesthesiologists, air traffic controllers, aircraft pilots, military personnel, car drivers and nuclear power plant operators have become any more or less important than they were in the 1950s (O'Brian and O'Hare, 2007). Perhaps it is more correct to say that the term complexity is not generally used in connection with things like children's games, mowing a lawn or buying groceries (although even here there is multiplicity of parts and degrees of interrelation, dynamism, uncertainty and difficulty). The point seems to be that, 'Something is complex if it contains a great deal of information that has a high utility, while something that contains a lot of useless or meaningless information is simply complicated' (Grand, 2000, p. 140).

Problem Spaces

Multiplicity, dynamism, uncertainty, difficulty, importance; these five factors represent one of two ways that the term complexity is used in the literature. In this case it is 'as a quality whose attributes […] characterise all complex systems' (Bar-Yam, 2004a, p. 3). This 'attribute view' of complexity, therefore, describes a five-dimensional 'problem space' with the curve in Figure 4.2 perhaps showing the trajectory of human factors through it. Of course, five dimensional problem spaces are not easy to visualise. Three dimensional spaces are, and it is possible to draw from the world of NEC research in order to illustrate an attempt at defining just such a thing. This is shown in Figure 4.2.

The overlap between the model shown in Figure 4.2 and the five attributes described above is largely coincidental, but nonetheless readily apparent. The *x*-axis is labelled familiarity, and is the extent to which facets of the problem space are well known. The proximal attribute is, of course, 'uncertainty', but 'multiplicity' and 'difficulty' clearly serve as distal causes. The *y*-axis is labelled rate of change, and of course this finds a ready link with 'dynamism'. The *z*-axis, labelled 'strength of information position' is defined as follows: 'Regardless of the degree of dynamism (though very possibly influenced by that factor) and the degree of knowledge available about it (though, again, very possibly influenced by it) the strength of the information position has an important impact on the applicability of [particular approaches to coping with complexity]' (Alberts and Hayes, 2006). In a sense this relates to 'importance', that some types of information about complex artefacts and entities are more important than others. Obviously,

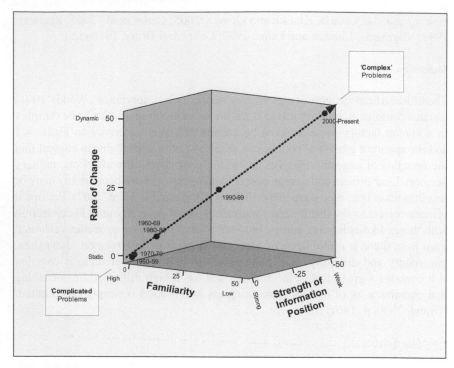

Figure 4.2   **Alberts and Hayes (2006) problem space provides a 3D approximation of the attribute view of complexity. The number of papers using 'complex' in their titles serves as a crude metric for the overarching nature of human factors problems and this is projected identically onto the *x, y* and *z* axes. A fit line is drawn through the coordinate points in order to provide an illustration of the general trajectory of human factors through this problem space**

this collapsed three dimensional version of the five-dimensional attribute view of complexity is not necessarily 'orthogonal or totally independent' (Alberts and Hayes, 2006, p. 76), neither are the five attributes just described regarded as final; furthermore, the problem space undeniably reflects a bias towards command and control issues, but, it still conveys a powerful point about the way ergonomic problems can differ from each other. Specifically, problems that are merely 'complicated' can be characterised by high familiarity (often bestowed by the adoption of a reductionist, atomic approach that decomposes phenomenon into small understandable units), non-dynamic rates of change (in which the human system interaction can be regarded as stable) and a strong information position (familiarity being 'very possibly' implicated in a strong information position). Complicated problems, therefore, find themselves in the bottom left-hand corner of the problem space (as shown in Figure 4.2).

Complex problems, on the other hand, are characterised by unfamiliarity (owing to the *interactions* between all those small, *not necessarily understandable* units), high rates of change (owing to the human system interaction being fundamentally *unstable*) and a weak information position (with familiarity and quantity of information *no longer* isomorphic with its importance). These types of problem occupy a position in the opposite corner of the problem space, the area into which the complexity curve seems to be pointing and which can be grounded in the set of key trends identified by Boehm (2006) in Table 4.1. These help to illustrate the kinds of 'real-world' challenges that human factors professionals are facing and why 'complex' arises as such an appropriate term for them.

The Complex Theoretic View

The second way that complexity is expressed in the literature, aside from an attribute, is to ascribe to it a 'quantitative measure, a single number that characterises a system,

Table 4.1   Wider trends associated with opposite corners of the human factors problem space (Boehm, 2006)

From complicated...	...to complex
A focus on specialisation.	An increasing integration of disciplines, specialisms and expertise.
A focus on what a system 'is' (i.e., requirements and functionality).	An increased emphasis on what a system 'does' (i.e., end value, effects and capabilities).
An increasing level of criticality and dependability required of *complicated* systems.	An increasing level of criticality and dependability required of *complex* systems.
A focus on constraining dynamism and imposing stable behaviour.	An increasing level of dynamism and rates of change.
A focus on stand-alone systems.	An increased emphasis on interoperability.
An emphasis on controlling complexity.	An increasing emphasis on ever more complex systems and systems of systems.
A focus on end-products often based on new technology which replaces obsolete equipment.	An increasing trend towards through life capability, integration of legacy systems and reuse.
An increase in computational power and the ability of entities and artefacts to exhibit *complicated* behaviour.	An increase in computational power and the ability of entities and artefacts to exhibit *complex* behaviour.

thus capturing the notion that some systems (mathematical or physical) are more or less complex than others' (Bar-Yam, 2004a, p. 3). This second viewpoint is that of 'Complexity Theory'. Here, complexity is equated to the amount of information needed to describe the phenomena under analysis in the Shannon (1948) information theoretic sense. The longer the description the more complex the phenomenon is, right up until the point at which complete randomness is reached when it 'cannot be described in shorter terms than by representing the [phenomenon] itself' (Bar-Yam, 1997).

Information Entropy

Under the rubric of information theory (e.g., Shannon, 1948), the amount of information needed to describe a system is based on the uncertainty inherent in that system. The greater the uncertainty the more difficult it becomes to predict future system states from current system states. Information entropy in these cases is said to be high. Each new state has to have more new information ascribed to it; thus the information content of its numerical description becomes longer, and it this that ultimately communicates something about the complexity of that system.

In cases where information entropy is low, when future states *can* be predicted from past states, then there is less need to encode each new state with more information (this information would be redundant). As a result, the information content of the system's description can be shorter for the same predictive efficiency. This, too, communicates something about the complexity of the system.

Closed systems tend to be associated with low information-entropy: new states do not contribute much additional information that can be used to describe the system numerically. Open systems (of various sorts) tend to be associated with high information-entropy: each new system state delivers new information to the system's description of itself. This entropic behaviour is reflected in various complexity metrics.

Implied in much of the extant NEC literature is high information-entropy and Moffat (2008) uses this approach to explore the issue much more thoroughly. From a human sciences point of view, the notion of problem spaces (i.e., Figure 4.1) is suggestive of parallelism, with NEC potentially able to exhibit the characteristics of open and dissipative systems in conjunction with more closed system states. It is likely that NEC's information entropy, and the subsequent behaviour of its complexity metrics, varies across time and function.

Kolmogorov Complexity

The theme of information entropy leads into the derivation of more specific complexity metrics, a notable core example being Kolmogorov complexity (Solomonoff, 1960). Underlying this is the principle of Computational Equivalence. The rationale behind this approach is that a computational model of a particular entity or artefact can be created which, if it were to be 'run', would generate and/ or fully explain the entity or artefact in question. Traditionally, such analyses are undertaken with computer programs serving as computational equivalents, but in human factors something like Hierarchical Task Analysis (HTA; Annett, 1971, 2005; Stanton, 2006) can serve an analogous purpose. After all, what is HTA if not a description of, and a computational equivalent for, its top-level goal? 'Running the HTA', performing all the operations, sub-goals and plans, should generate the top-level goal, and because of this, it has computational equivalence to the actual entity or artefact which in real life produces it. Complexity theory of this Kolmogorov variety is about subjecting models like this to analysis in order to diagnose the 'real' entity or phenomenon'ss complexity from its computational surrogate, whether it be a computer program or in this case an HTA.

Complexity Metrics

Some examples of metrics derived from complexity theory, that is to say the quantitative measures, the single numbers that can characterise a system, are applied to HTA for illustration and detailed in Table 4.2. At this point it is relevant to note that HTA is not the only computational equivalent available to human factors practitioners. We have applied several of these metrics successfully to social networks, and there is good reason to suspect that they can provide a measure of complexity for all manner of other representations, from process diagrams to fault trees. For the time being, though, Table 4.2 focuses on HTA.

Table 4.2   Examples of complexity theory metrics (from Hornby, 2007) explained using Hierarchical Task Analysis as the 'computational equivalent' for an entity or artefact

Name	Description
Number of build symbols	The number of goals/sub-goals that produce the overall goal
Algorithmic complexity/ algorithmic information content/Kolmogorov metric	The number of goals and plans that produce the overall goal
Logical depth	The *minimum* number of sub-goals and plans that will produce the overall goal
Sophistication	The total number of logical operators (e.g., IF, THEN, AND, OR) that are used in the plans in order to generate the overall goal
Grammar size	The number of new and distinct logical operators used to generate the overall goal
Connectivity	The maximum number of edges that can be removed before the task analysis splits into two
Height	The maximum number of links between sub-goals from the bottom to the top of the analysis

Interestingly, the complex theoretic approach does have a minor human-factors legacy, albeit limited to the specialist realm of axiomatic design (e.g., Suh, 2007). This quantitative approach to design lends itself well to similarly quantitative measures of complexity. Unfortunately, complexity metrics are employed that are so specific to this approach (for example, complexity is seen 'as a measure of uncertainty in achieving the specified' [Functional Requirements]; Suh, 2007, p. 111) that its lack of wider usage is hardly surprising.

Axiomatic design reveals a further, somewhat wider limitation. If an equilibrium state is defined as a physical design that has arisen from mathematical axioms, or an overall goal that has been generated from an HTA, then in both these cases this state has been predicated on strongly rational assumptions (Burke, Fournier and Prasad, 2006), assumptions which focus on stable behaviour, well-understood dynamics and a form of 'IF THEN' logic (i.e., a set of pre-requisites that are firmly located in the 'complicated' octant of the problem space shown above in Figure 4.2). A common criticism of the complex theoretic approach, therefore, is that mathematical proofs like this are able to describe rational, and by implication simple, or at least merely 'complicated' systems, but only certain aspects of truly complex ones (Boguta, 2005, p. 18). For analogous 'non-equilibrium' systems, ones that embody various degrees of change, uncertainty and throughput, as is the case for most sociotechnical systems, we need to go beyond both the attribute view and the complex theoretic view to a further approach, one that is encapsulated under the heading 'complex systems research' and which forms the backdrop for the remainder of this chapter, and indeed, the rest of the book.

The Complex Systems Research View

Components of the 'attribute' and 'complex theory' views are highly useful, yet they still do not, on their own, completely get to the heart of what it means for the real-world problem spaces increasingly occupied by human factors. Here we are talking about 'emergent behaviour exhibited by interacting systems operating at the threshold of stability and chaos' (Roetzheim, 2007, p. 4), of 'systems with a large number of interacting parts and a large throughput of energy, information, or material (Hubler, 2007, p. 10; Hubler, 2006; Hubler, 2005) and systems which 'don't just passively respond to events [...] they actively try to turn whatever happens to their advantage', which is to say they are 'adaptive' (Waldrop, 1992, p. 11).

The field variously called 'Complex Systems', 'Complex Adaptive Systems' and/or 'Complex Systems Research' would recognise this as the essence of what true complexity, as a distinct phenomenon, is all about. Highlighting the fluid and complex nature of 'complexity' itself, we at this point run into a problem of terminology. Some authors (e.g., Moffat, 2003) use what we have called 'Complexity Theory' (above) to describe what we have just attempted to label 'Complex Systems'. They (and we) are both correct. For the purposes of this human-factors based exploration of the concept we use Complex Systems in the same way that Moffat uses Complexity Theory, 'as a shorthand term to cover a number of areas, each with its own distinct heritage' (Moffat, 2003). If the attribute view of complexity is our first view, the complexity theory view, that of numerical metrics, is our second view. The complex systems research view is our third.

Regardless of the label given to it, complexity research of this third type is concerned with the kinds of problems that 'emergent behaviours at the boundary of stability and chaos' create. Appropriately termed 'wicked problems' by systems engineers, these are less amenable to the 'complex theory' approach as they are often not fully understood, have fuzzy boundaries, lots of stakeholders and lots of constraints with no clear solution (Rittel and Webber, 1973). They are also not completely explained by the 'attribute' view of complexity, which is better at answering the 'what' of complexity rather than the underlying 'how' and 'why'. Truly complex problems 'are the problems that persist – the problems that bounce back and continue to haunt us' (Bar-Yam, 2004a, p. 14). Ironically, they also tend to be the very problems that prompt the intervention of last resort: the human factors professional.

Of course, problems of a 'non-wicked' nature are what the stated regime is designed to yield, and in most cases 'does' yield. But to the extent that 'wicked problems' nonetheless still appear is what makes complex systems research highly relevant to human factors. Wicked problems 'may appear to be scattered examples from the corner of [disparate scientific] fields. But in fact they are just the borders to a vast world that science has assiduously evolved to avoid' (Boguta, 2005, p. 16).

Atomic Methods for an Atomic Age

The reason that science, and as we shall argue, human factors, have appeared to assiduously avoid this third type of complexity is due in large part to the scientific methods employed. Complex-systems research advocates an inversion to what it sees as the reductionist, decompositional, Newtonian logic of classical thinking (e.g., Waldrup, 1992; Bar-Yam, 2004a). Instead of a top-down approach to investigating problems, complex-systems research 'studies how relationships between parts give rise to the collective behaviours of a system and how the system forms relationships with its environment' (Bar-Yam, 2004a, p. 24). In numerous fields where this approach has been tried, which include extremely large-scale examples of collective behaviour (such as global weather patterns and social systems the size of nations), all are understood as phenomena that are not adequately explained by understanding their constituent parts. Instead, what is required is a focus on the 'relationships' between those parts, relationships that are normally fractured and discarded by the reductionist method of problem-solving. In human factors, concepts and methods in human error and cognitive-systems engineering, for example, Reason's layered systems 'Swiss cheese' model (Reason, 1990), Perrow's 'normal accidents' (1999), Hollnagel and Woods' 'joint cognitive systems' (2005) and Vicente's Cognitive Work Analysis (1999) seem to come closest to not just dealing with complexity, but specifically 'this' type of complexity, although rarely is it defined explicitly in this way. Complex-systems research presents a much more forceful agenda than either of these human factors specialisms in terms of challenging the ubiquity of the reductionist method. Whilst representing the touchstone of scientific progress to date it is argued that it is in danger of not just becoming an impediment to future progress, but an optimum strategy for ensuring its failure (Berman, 1983; Bar-Yam, 2004a).

Physics Envy

A very crude characterization of the so-called 'classic approach' to science is that in order to understand how something works and behaves it has to be taken apart. This has been the basis of scientific endeavour and progress since the Renaissance and Carl Linnaeus, the acknowledged father of taxonomy, produced such a thing in 1735 for plant and animal species and called it the 'Systema Naturae'. The rationale behind taking things apart and reducing them to their fundamental properties is that it makes them less complex and easier to understand. When all parts have been understood, they can be reassembled into the 'whole' from whence they came using the hierarchical map of decomposition that was used to take the system apart in the first place. The whole, therefore, is assumed to be no more and no less than the sum of its parts. It is precisely by these means that true complexity, in the complex systems research sense of the term, 'can be assiduously avoided'.

Emblematic of the progress and success of this approach is physics. Taxonomy and reductionism might have had a longer history in the field of biology, but physics can lay claim to the finest level of decomposition (in the form of particles smaller than atoms) and the most erudite set of fundamental laws and theories that explain them, stretching forward in time from Newton to Einstein. No wonder, then, that 'scientists in disciplines outside physics wished their own subjects could boast the intellectual profundity, the mathematical agility and the foundational rigour that was evident in physics' (Ball, 2005, p. 256). This phenomenon is light-heartedly referred to as 'physics envy' and lest there be any doubt that human factors is not afflicted with it (on occasion at least), then consider for a moment Venda's paper (1995) in which first, second and third laws of so-called 'ergodynamics' are expounded. Despite wearing the clothes of mathematics, human factors, like any human science, cannot hope to attain anything approaching the same level of precision (Ball, 2005). Because of this, things like ergodynamics fall foul of the same problems that complex theoretic approaches do: they only deal with certain aspects of complexity, and not those described by the complex-systems research perspective. Nevertheless, the influence of 'physics envy' (to persist with the colloquialism) is much more implicit and pervasive than the highly specialist nature of 'theories of ergonomics' and 'axiomatic design' papers might otherwise suggest. In the widespread science and practice of human factors, the simple reductionist expedients of decomposition and hierarchy are alive and well.

Consider for a moment what could be regarded as a fundamental unit of human factors analysis. Physics has the atom, human factors has 'the task': 'something that needs to be done, an act that one must accomplish' (Reber, 1995, p. 784). To judge by its long legacy and enduring popularity among practitioners, one of the central analysis methodologies in human factors is undoubtedly Task Analysis which, generally speaking, involves identifying tasks, collecting task data, analysing the data so that tasks are understood, and then producing a documented representation of the analysed tasks (Annett et al., 1971). Armed with this data on tasks, human-factors practitioners can then begin to answer useful questions related to them (Stanton, 2006), such as 'What is the workload associated with this task?' Who should perform that task? What are the situational awareness requirements? What are the error probabilities? And so on. As Table 4.3 shows, all of the half dozen task analysis methods in widespread use today rely to some extent on taking tasks apart (i.e., decomposition), drawing a map of what, and how the task has been dismantled (i.e., by creating a hierarchy) and analysing the pieces in fine detail (i.e., exhaustive description and re-description).

Table 4.3   Task analysis methods all rely on the analysis of parts in order to understand the whole (data drawn from Stanton et al., 2005)

Name/acronym	'Whole' under analysis	Brief description
Hierarchical Task Analysis (HTA)	Goals	Describes 'activity under analysis in terms of a **hierarchy** of goals, sub-goals, operations and plans' (p. 46)
Goals, operators, methods and selection rules (GOMS)	Human computer interaction	'GOMS attempts to define user goals, **decompose** those goals into sub-goals and demonstrate how the goals are achieved through user interaction' (p. 54)
Verbal protocol analysis (VPA)	Processes, cognitive and physical, used to perform a task	'It is recommended that a HTA [**Hierarchical** Task Analysis] is used to describe the task under analysis' (p. 58)
Task decomposition	Task or scenario	'...using specific task-related information to **decompose** the task in terms of specific statements' (p. 62)
Sub-goal template method (SGT)	Information requirements for tasks	'...involves **re-describing** a HTA for the task(s) under analysis' (p. 68)
Tabular task analysis (TTA)	Task or scenario	'...**takes each bottom level task step** from a HTA and analyses specific aspects' (p. 72)

Out of a total of 91 human factors methods that are widely available, readily applicable, original, and in widespread use (Stanton et al., 2005), 65 per cent of them either rely explicitly on these task analysis methods or else some other form of decomposition, hierarchy and/or exhaustive re-description. Not all methods and approaches are like this, but clearly, taking things apart in order to understand them is an undeniable hallmark of a significant tranche of human-factors practice. Of course, it goes without saying that significant and worthwhile progress has been made using it. But as a strategy for understanding complex phenomena, not just those that are comprised of lots of parts and difficulty, or those that can have numerical values ascribed to them, but phenomena at the edge of stability and chaos, the limitations of this reductionist approach are starting to become apparent.

In physics, understanding nature's laws on these terms leaves 'unanswered the question of how to apply those laws to any but the simplest of systems' (Gleick, 1987). In human factors, it 'does not offer many of the concepts and techniques needed' (Bainbridge, 1993, p. 1399). De Greene (1980) looks at 'major conceptual problems in the systems management of ergonomics research' and notes the difficulties that stem from 'future uncertainty, the use of static models, the relationships between models and data' and so on (p. 3). These traditional methods find themselves increasingly criticised because they seem to 'under-estimate (if they capture them at all) the context-dependent aspects of human performance', i.e., the complex ones (McLeod, Walker and Moray, 2005, p. 673). The question here is not one of a strict systems-versus-classical dichotomy, in which each is completely orthogonal; the very notion of a problem 'space' (however conjectural) highlights instead the idea of transitive complexity. Problems can exhibit a high degree of parallelism and overlap, exhibiting traits that are amenable to classical, bottom-up, reductionist approaches simultaneous with traits that require a systemic, top-down approach.

Command and Control

One problem domain that has played a prominent role in accelerating the use of terms like 'complex' and 'complexity' in the titles of papers published in the human factors literature is command and control. This is because 'the relationship between complexity and 'information-based' warfare is […] less deterministic and more emergent; less focussed on the physical, and more behavioural; less focussed on things, and more on relationships.' (Moffat, 2003, p. 3). It is also because the 'post-9/11 world suggests a different set of mechanics for competition and conflict, a new model of conflict, and different operational and tactical problems from those upon which we have grown to focus' (Smith, 2006, p. 5). Command and control is undeniably complex and, if anything, is becoming more so. This chapter is certainly not the first time that command and control has been examined through this complexity lens; indeed, it is a mainstay of NEC research. Whilst it is not possible to be comprehensive, what we have set out to achieve in this section is a further harvest of complexity concepts. The aim is to start using them as a way of providing insight into perplexing human factors phenomena.

Complexity and Scale

If hierarchies are a generic approach to science then they are certainly a generic approach to organisational design (Davis, 1977; Toffler, 1981). With reference to the NATO SAS-050 approach space (the model of command and control seen in Chapter 2), so-called classic C2 can be seen as having unitary decision rights, tightly constrained patterns of interaction and tight control over the distribution of information (e.g., Alberts and Hayes, 2006). Aside from the rationalistic principles of efficiency, predictability, quantification and control that this position in the approach space inplies, it is possible to say a number of further things from a complex-systems research point of view. The first of these relates to the idea of 'scale', a concept developed by researchers at the New England Complex Systems Institute (e.g., Bar-Yam 2004a, b). They state that there is a trade off between scale of observation, defined as the 'level of detail visible to an observer of a system'; Bar-Yam, 2002, p. 1) and the sorts of behaviours occurring at different levels (more specifically the number of 'distinct' behaviours occurring). Depending on how an organisation is designed, this behaviour changes in quite specific ways depending upon the scale of observation. This relationship is termed a complexity profile and defined as 'the amount of information necessary to describe a system as a function of the level of detail provided' (Bar-Yam, 2002, p. 1). In the case of classic C2 its primary behaviours are generally visible at large scales. Consider the example of ancient armies marching en masse. The reason it is visible at this large scale of observation is because the entities and actors are behaving in highly coordinated ways or, to use Perrow's terminology, the system is 'tightly coupled' (1999). This arises as a direct result of the hierarchical management infrastructure.

As we zoom in on this stereotypical organisation, decreasing our scale of observation to that of small groups of people, that high level of coordination gives rise to a distinctive property. Although the effects of hierarchical command and control can be viewed from a large scale of observation, its fine scale behaviour is not always especially complex. It is an example of a complex organisation (in terms of its control structures, rules, myriad procedures and patterns of vertical communication) which nonetheless only really permits people to undertake simple tasks (e.g., Sitter, Hertog and Dankbaar, 1997). This, according to Bar-Yam (2004b), reflects a fundamental principle of complex, and not so complex, systems: when parts of a system are acting together, the fine scale complexity is small.

The second point about hierarchies is that the large-scale behaviours they are capable of cannot be more complex than the person(s) at the top of the structure, which is complex but ultimately limited (Bar-Yam, 2004b). Hierarchies are large scale but also low variety. Variety is the name for a formal concept in cybernetics which refers to the total number of states that a system can adopt, or the number of behaviours it can emit, or its degrees of freedom (Ashby, 1956). In discussions of scale, it is variety that is often being used as shorthand for complexity. Less complex is equated with only limited numbers of behaviours (low variety), and vice versa for more complex.

Ashby's Law of Requisite Variety is a cybernetic principle founded on a raft of concepts too numerous to present here (the reader is referred to Ashby, 1956, for a classic introduction to the topic). The Law of Requisite Variety states, quite simply, that the degrees of freedom/behaviours/states that a system is capable of must match the degrees of freedom/behaviour/states that the environment within which the system resides is capable of; or to put it another way:

> 'in active regulation only variety can destroy variety. It leads to the somewhat
> counterintuitive observation that the regulator must have a sufficiently large
> variety of actions in order to ensure a sufficiently small variety of outcomes
> in the essential variables [.] This principle has important implications for
> practical situations: since the variety of perturbations a system can potentially
> be confronted with is unlimited, we should always try to maximize its internal
> variety (or diversity), so as to be optimally prepared for any foreseeable or
> unforeseeable contingency.' (Heylighen and Joslyn, 2001)

As Ashby states, if 'the system is continuous, we can ask whether it is stable against all disturbances within a certain range of values' (1956, p. 85). However, the idea of 'stable' is problematic and the Law of Requisite Variety brings some of these issues into relief. Ashby saw stability as a property which in most useful cases could only be specified over a brief time envelope and only then under certain conditions:

> 'As shorthand, when the phenomena are suitably simple, words such as
> equilibrium and stability are of great value [but] phenomena will not always
> have the simplicity that these words presuppose.' (Ashby, 1956, p. 85)

Whilst Ashby recognised that certain aspects of systems were invariant, he also recognised evolution, which is something that Requisite Variety connotes quite strongly in some cases. In possessing Requisite Variety, a system doesn't just posses the means to move itself from one state to another, but in some cases to use those states to inform future states and thereby alter the invariant aspects of the system itself.

Referring back to the problem space (Figure 4.2) that was derived from the 'attribute view' of complexity, clearly the area of the space bounded by high familiarity, static rates of change and a strong information position connotes low variety. In this situation:

> 'Control or regulation is most fundamentally formulated as a reduction of variety:
> perturbations with high variety affect the system's internal state, which should

be kept as close as possible to the goal state, and therefore exhibit a low variety. So in a sense control prevents the transmission of variety from environment to system.' (Heylighen and Joslyn, 2001)

Such an organisation would be something akin to classic C2. But what happens when the area of the problem space being occupied by classic C2 shifts, as is evidently the case, towards the high variety that arises from a weak information position, unfamiliarity and dynamism? Whilst '...the hierarchy is good at amplifying, increasing the scale of behaviour of an individual', and by doing so meeting the comparatively low-variety requirements of one octant of the problem space, it is 'not able to provide a system with larger complexity than that of its parts'. As a result, variety does indeed start to destroy variety and the organisation starts to become afflicted with the performance sapping pathologies characteristic of all bureaucratic organisations (Ritzer, 1993).

Matching Approaches to Problems

The conceptual response to the challenges present in alternative, more complex, areas of the problem space is NEC. NEC can, in theory, exhibit greater variety by being based on a non-hierarchical structure, one that emphasises smaller semi-autonomous groups and minimum critical specification. As shown in Chapter 2, NEC can be characterised by distributed patterns of interaction, peer-to-peer decision rights and broad dissemination of information. This means that it occupies the diagonally opposite region in the NATO SAS-050 command and control approach space to that occupied by classic C2. At a more fundamental level this type of command and control can be seen as an attempt to maximise a different set of performance characteristics, these being:

- Instead of efficiency and amplifying the effect of the few individuals who reside at the top of a hierarchy, emphasis is given to self-synchronisation or '...forces that can work together to adapt to a changing environment'.
- Instead of predictability and coordinated, coherent action, there is emphasis on shared awareness in order 'to develop a shared view of how best to employ force and effect to defeat the enemy'.
- Instead of quantification and organisational reductionism, there is a shift from 'doing things better' to 'doing better things' entirely.
- Finally, instead of control and nearly automatic functioning bounded by rules and structures, NEC 'removes traditional command hierarchies and empowers individual units to interpret the broad command intent and evolve a flexible execution strategy with their peers' (Ferbrache, 2005, p. 104).

As before, two further points about scale and variety can be made. In terms of scale, the reverse situation to hierarchical command and control pertains to

NEC. Its behaviour is not necessarily as visible at course scales of observation (as is the case with classic C2); however, as the scale of observation is decreased complexity (or variety) increases. The reason for this is that finer scales reveal not echelons of actors and entities performing simplified, repeated tasks en masse, but self-synchronising teams who have much greater freedom of action. This also is a fundamental principle of scale and complexity: 'When parts are acting independently, the fine scale behaviour is more complex' (Bar-Yam, 2004a, p. 58). The complexity profile that this scale dependent level of variety traces is thus completely different to the comparable profile traced by classic C2 as Figure 4.3 shows.

The second point to make here is that unlike hierarchical command and control, NEC's behaviour *can* be more complex than the person(s) at the top of the structure. The focus on self-synchronising teams, effects-based operations, shared awareness and so forth, creates the conditions for greater freedom of action, that is to say more variety. If it is only variety that can destroy variety, then NEC

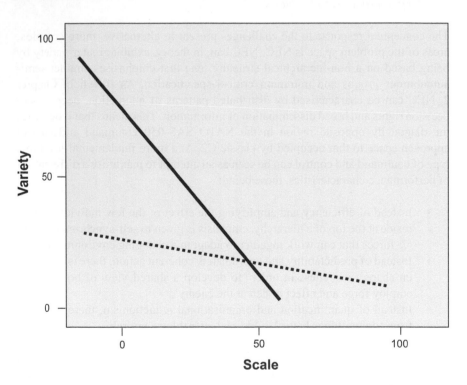

Figure 4.3   Complexity profiles for NEC and traditional hierarchical command and control. The profiles arise from considering the number of distinct behaviours (i.e., the variety) available to the organisation at different levels of observation (i.e., scale)

Source: Bar-Yam, 2003b.

should be more robust when partnered to areas of the problem space bounded by unfamiliarity, dynamism and a weak information position. This would certainly appear to be the case based on the two previous chapters.

Naturally, a degree of caricaturing and stereotyping has been necessary to draw out these key differences between NEC and classic C2, so in order to temper any over-enthusiasm for one approach or another it is important, firstly, to state that scale versus complexity is a trade-off. There are situations which require sheer scale, and others that require high organisational variety. As stated before, the very notion of a problem and approach 'space' highlights the fact that important aspects of complexity (and the response to it) vary as a function of time and of behaviour; they are not orthogonal. That being the case, it behoves organisations to try to match their 'approach' to the extant 'problem', as Table 4.4 shows. When a so-called 'wicked problem' meets a C2 'approach', the mismatch in variety (or lack thereof) results in irrational system behaviour or something approximating to chaos (e.g., Gleick, 1987). When an NEC approach meets a stable problem there is also a potential mismatch. The increased variety is redundant and the relative lack of coordination means that the ultimate scale of action will be limited. On the other hand, when a C2 approach meets a stable problem there is a match and rational, deterministic, machine-like behaviour can arise. Likewise, when an NEC approach meets a wicked problem there is also a match, the organisation has equal variety to the context in which it is operating and complex behaviour can 'emerge'.

Table 4.4   Matrix of 'Approach' versus 'Problem' and a simple taxonomy of resultant system behaviours

		Problem	
		Deterministic	Wicked
Approach	C2	Rational	Chaotic
	NEC	Redundant	Complex

This discussion serves as a powerful metaphor for human factors. No one is denying the achievements that have been made using existing approaches (reductionist or otherwise) or that can continue to be made by using them. It is also fair to say that a putative human-factors problem possesses a high degree of parallelism or so-called transitive complexity, simultaneously occupying a part of the problem space for which deterministic, reductionist approaches are entirely appropriate. The question to ask, however, is where the complexity curve is pointing? As in the case of NEC, it behoves human factors to attempt also to match its approaches to parts of the problem space being occupied by putative human-factors problems. It is not a question of being one thing or another, reductionism versus complexity, but a complementary approach which recognises the limitations inherent in both.

Summary

Complexity is about there being several parts – about difficulty, dynamism, interconnection, change, uncertainty, fuzziness, teamworking, information throughput, the equifinal expedient of their being multiple ways to achieve multiple end states, of evolution and of adaptability. It is also about emergent properties, phase spaces, critical points, transitions, chaos, bifurcations and sensitive dependence on initial conditions. It is also related to algorithmic complexity, logical depth, computational equivalence, cybernetics, variety (requisite or otherwise) and information theory. Above all, complexity is itself complex. Associated with each of the myriad terms is considerable conceptual baggage which in this chapter has either been ignored (as we focus on breadth not depth) or else simplified (as we sacrifice strict scientific rigour for analogy and innovation). In this chapter we have merely tried to establish a conduit for knowledge by distilling the multifarious nature of complexity through three overlapping views:

1. The attribute view: here complexity is about the multidimensional problem space through which human factors is travelling. Here we showed how this multi-dimensional space could be collapsed into three critical variables in order to create a defined 'problem space'.
2. The complex theoretic view: which deals with the quantification of certain aspects of complexity. Here we showed how the metrics provided by this approach readily complement existing mainstays in human factors methodology, allowing a better judgement to be made about whether one entity or phenomena is numerically more complex than another.
3. Finally, and above all else, there is the complex-systems research view. The domain of 'true complexity', of the emergent phenomena occurring at the edge of chaos, and the phenomena which require a new set of tools and approaches in order to diagnose what is going on and what to do about it. Here, complexity has been related to the 'approach' that is needed to match putative 'problems'. Complex problems (i.e., those that are unfamiliar/ unstable/unknowable) require systems to be configured in certain ways (i.e., peer-to-peer interaction, devolved decision rights and widespread dissemination of information). Complex problems also require human factors itself to be configured in certain ways, for its various 'approaches' to match its various 'problems'. It is to this that we can now turn. Armed with the triad of socio, technical and complex we can now progress into the analysis of a large-scale example of 'real-life' NEC. There is, however, more to be said about complexity and the topic will be returned to again in Chapter 6.

Chapter 5
Dimensions of Live-NEC

Aims of the Chapter

This chapter takes the NATO SAS-050 Approach Space, a widely publicised model of command and control, and gives each of its primary axes a quantitative measure using social network analysis. Deriving such measures means that the actual point in the approach space adopted by real-life command and control organisations can be plotted, along with the way in which that point varies over time and function. The chapter is divided into two parts. The first part presents the rationale behind this innovation and how the newly extended NATO SAS-050 approach space was subject to verification using theoretical data. The second part applies these insights to a large scale military command post exercise. The main findings are based on the specific region in the approach space live-NEC actually occupied (rather than the region it thought it might occupy), the extent to which this 'approach' matched the extant nature of the 'problem' and the dynamics of the organisation in terms of agility and tempo. Novel ways of representing these twin concepts are illustrated. This chapter is purely exploratory and deals principally with the 'what' of live-NEC. The next two chapters go into more detail about the 'why'.

Introduction

Background and Context

If NEC's theoretical vision is sometimes blurred and ill-defined, then so too is the real-life instantiation of it. Practical realisations of NEC range from something akin to a computer network at one end of the spectrum (the inspiration provided by Wal Mart's vertically integrated systems has already been mentioned; e.g., Shachtman, 2007) to an organisation that exhibits the behaviour of an organism at the other (in which the inspiration derives, albeit indirectly, from the legacy of sociotechnical systems). Not surprisingly, the practical realisation of NEC adopts different positions along this continuum with widely different forms of organisational infrastructure, some of it ostensibly 'technical' in nature, combined with operational procedures that vary between the scripting of tasks and micro-management through to a focus on semi-autonomous groups and so-called effects-based operations. Because of this, the state of the art in terms of the real-world implementation of NEC is currently, and perhaps will always remain, something that is much more akin to a process than it is to a fixed end-state (Alberts and

Hayes, 2003). In other words, it is difficult to point to any one version of NEC and claim 'that's it'.

The NATO SAS-050 Approach Space

If command and control, in its most generic state, can be regarded as simply the management infrastructure for any large, complex, dynamic resource system (Harris and White, 1987) then clearly not all 'management infrastructures' are alike. In the case of NEC the label 'command and control' may not always be very helpful (e.g., Alberts, 2007). The NATO SAS-050 approach space shows this to be the case. When the formal definitions of command and control are brought to bear, as in the highly stereotypical case of so-called 'classic C2' the management infrastructure that results occupies only a relatively small area of the approach space compared to the various incipient states of NEC described above, which would seem to be widely scattered. The NATO SAS-050 approach space represents an important reality of C2 and is a well-publicised basis for exploring and investigating it (e.g., NATO, 2007; Alberts and Hayes, 2006; Alberts, 2007).

The model is a research output of a NATO working group. Its development is reasonably well rehearsed in the literature (see NATO, 2007 and Alberts and Hayes, 2006). It derives from an underlying information processing view of command and control expressed as a generic C2 'approach' (a cyclical pattern of information collection, sense making and actions). This in turn enabled a comprehensive reference model to be developed, comprised of over 300 variables that map onto this approach. These variables were drawn from fields as diverse as general systems theory, human factors, cognitive psychology and operational research (amongst many others). The 300 variables were then connected by over 3000 links and various systems engineering methods applied in order to undertake a process of 'dimension reduction'. By this process, what are termed 'three key factors that define the essence of [command and control]' (Alberts and Hayes, 2006, p. 74) are arrived at. These are:

x = allocation of decision rights (from unitary to peer-to-peer),
y = patterns of interaction (from fully hierarchical to fully distributed),
z = distribution of information (from tight control to broad dissemination).

These key factors form intersecting x, y and z axes and the three dimensional NATO SAS-050 approach space in its graphical form (e.g., as shown in Chapter 2, Figure 2.1). In theory, a command and control organisation can be positioned along its respective x, y and z axes and its location in the approach space fixed. It is possible to go further than this. Because NEC is contingent (e.g., Mintzberg, 1979) upon its environment, which is to say that it behaves as a dynamical system, the changing values of x, y and z over time give rise to successive points in the three-dimensional space. When linked together they form a trajectory, an abstract representation of motion that describes the dynamic behaviour of a complex system like NEC. Within the field of complex systems research the practice of dimension reduction and the creation of simple coordinate spaces into which a real-life dynamical system can be plotted is referred to not as an 'approach space' but a 'phase space' (Gleick, 1987). We will deal with the NATO SAS-050's retrospective legacy in complex systems research in the next chapter. For the time being it can be noted that the more extensive the motion within the approach space the greater the organisation's 'variety' is said to be, and thus, according to the Law of Requisite Variety, the organisation's ability to cope with complexity (Ashby, 1956). In the language of C2, as Alberts and Hayes (2006) point out, the greater the number of points in the approach space that can be occupied, then the greater the command and control organisation's agility (Alberts and Hayes, 2006).

What drives this motion through the approach/phase space? What causes command and control organisations to change in terms of the three fundamental dimensions? According to NATO (2006) the two proximal sources of these dynamics are function (in that different parts of a command and control organisation are doing different things) and time (the organisation configures itself differently at different points in an evolving situation). Both of these dynamics, in turn, are dictated by the distal source of dynamics represented by the causal texture of the environment or problem that the command and control organisation is operating under, and which in turn it is influencing (Emery and Trist, 1965). The NATO approach space is therefore joined by a corresponding 'Problem Space'. To briefly recap, this too has three dimensions that form intersecting axes and a three dimensional space. These are as follows:

x = familiarity (from high to low),
y = rate of change (from static to dynamic),
z = strength of information position (from strong to weak).

As described in the last chapter, problems that are merely 'complicated' can be characterised by high familiarity (of underlying principles), non-dynamic rates of change (the situation is stable) and consequently a strong information position. The type of problem to which NEC is a conceptual response is arguably 'complex' rather than merely 'complicated'. Complex problems are the ones that can be characterised by unfamiliarity, change and a weak information position. The approach and problem spaces are yoked, as shown in Figure 5.1. The area occupied by a putative 'problem' in one, needs to be matched by an organisation occupying a corresponding area of the 'approach' in the other. If agility describes this match in terms of the area or regions in the approach space occupied by an organisation then the ability for it to track the dynamics of the extant problem in a timely manner can be referred to as tempo. Appropriate levels of agility and tempo mean that the variety or degrees of freedom offered by the problem are able to not just be matched, but also tracked by the variety and/or agility inherent in the command and control organisation (Bar-Yam, 2004; Alberts and Hayes, 2006).

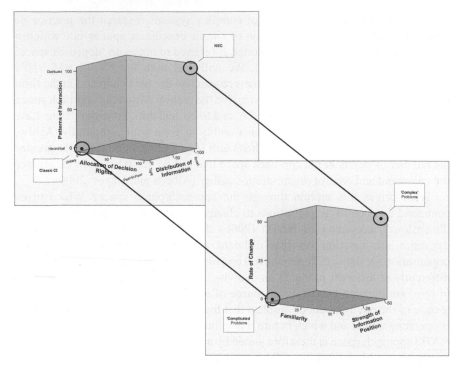

Figure 5.1   The NATO SAS-050 approach space is yoked to a corresponding problem space

By fixing and tracking different approaches to command and control, as well as the problems to which they are directed, the ultimate aim of the approach and problem spaces can be met, which is to facilitate exploration of 'new, network enabled approaches […] to command and control and compare their characteristics, performance, effectiveness, and agility to traditional approaches to command and control' (NATO, 2007, p. 7).

The Missing Links

If the NATO SAS-050 approach space 'is intended to serve as a point of departure for researchers, analysts, and experimenters engaging in C2-related research' (NATO, 2007, p. 3) then it is legitimate to extend the model. Specifically, to extend it in such ways that it speaks towards the following three aspirations:

1. 'We are interested in the actual place or region in this [approach] space where an organisation operates, not where they think they are or where they formally place themselves' (Alberts and Hayes, 2006, p. 75). The requirement that flows out of this is for metrics to define, quantitatively, the position that live command and control organisations adopt on any one of

the approach space's three axes. If live command and control can be fixed into the approach space, then it can be compared 'to traditional approaches to command and control' (or indeed to any other similar organisation).

2. The observation that 'an organisation's location in the C2 approach space usually ranges across both function and time' (p. 76) brings a further requirement to capture and understand the underlying dynamics of NEC. If the underlying dynamics can be captured and understood, then useful aspects of tempo and agility can be revealed.

3. Finally, 'Identifying the crucial elements of the problem space and matching regions in this space to regions in the C2 approach is a high priority'. Fixing and understanding the dynamics of command and control in the approach space increases the accuracy of the mapping that can occur between approach and problem, thus moving judgments about efficacy and performance from relative towards absolute.

The first part of this chapter deals with the innovations that enable the NATO SAS-050 approach space to be transformed from a typology into a taxonomy and thus help to meet the three objectives above. An explicit strategy for achieving this is derived from social network analysis and is put to the test with theoretical data. Testing the hybrid social network/NATO SAS-050 approach space with live data in the second part provides an opportunity to observe 'the actual place or region where an organisation operates', how that location varies 'according to function and time' and, aided by theoretical data, 'match regions of the problem space to the approach space'.

Part 1: Developing the NATO SAS-050 Model

Social Network Analysis

At the heart of all the missing links presented above is the ability to provide quantitative measures, or metrics, that relate meaningfully to decision rights, patterns of interaction and dissemination of information. Social Network Analysis (SNA) is used to overcome this limitation.

In general terms a social network is 'a set of entities and actors [...] who have some type of relationship with one another.' Social network 'analysis' represents 'a method for analyzing relationships between social entities' (Driskell and Mullen, 2005, p. 58-1). A social network is created by plotting who is communicating with whom on a grid-like matrix. The entries into this grid denote the presence, direction and frequency of a communication. The matrix can be populated using information drawn from organisation charts and standard operating procedures so that it describes what communications structure *should* occur. Much more consistent with the approach space, however, is that the matrix can also be populated with live data which describes what communications structures *actually* occur.

The matrix of agents and links is what enables a social network diagram to be created. This is a graphical representation of the entities and actors who are linked together where, obviously, apart from very simplistic networks any underlying patterns extant in this cobweb of nodes and links is normally very difficult to discern by eye alone. Thus, graph theoretic methods are applied to the matrix in order to derive a number of specific social network metrics (e.g., Harary, 1994). These metrics form the basis of a comprehensive diagnosis of the network's underlying properties, which include several which seem to relate to decision rights, patterns of interaction and distribution of information. This mapping of social network metrics to the NATO SAS-050 model axes is a key innovation and is described below.

Decision Rights Mapped to Sociometric Status

Decision rights can be mapped to the social network metric called 'B-L Centrality' (B and L referring to the originators, Bevelas and Leavitt). B-L Centrality is given by the formula:

$$B\text{-}LCentrality = \frac{\sum_{i}^{g} = 1; j = 1\delta_{ij}}{\sum_{j=1}^{g}(\delta_{ij} + \delta_{ji})}$$

B-L Centrality states that 'The most central position in a pattern is the position closest to all other positions' (Leavitt, 1951, p. 38). g is the size of the network and is equal to the number of nodes and links (or vertices and edges in social network language). δ_{ji} is the shortest path between two nodes (i and j) or geodesic distance. The hypothesis, therefore, is that hierarchical networks would generally posses fewer highly central agents (corresponding to unitary decision rights) compared to peer-to-peer networks. Specifically, the number of agents scoring more than the mean centrality score for a given network will be higher for something like NEC because it embodies greater extents of peer-to-peer interaction than is the case for classic C2, where only a few key nodes have a high ability to communicate.

Patterns of Interaction Mapped to Network Diameter

Patterns of interaction can be mapped to the social network metric 'diameter', which is given by the formula:

$$Diameter = \max_{uy} d(u, v)$$

where d(u, v) is 'the largest number of [agents] which must be traversed in order to travel from one [agent] to another when paths which backtrack, detour, or loop are excluded from consideration' (i.e., $\max_{uy}$; Weisstein, 2008; Harary, 1994). Generally speaking, the bigger the diameter, the more agents there are on lines of communication. The hypothesis is that a peer-to-peer organisation facilitates more direct and therefore distributed communication (and thus has a smaller diameter) than a hierarchical network, with more intermediate layers in between sender and receiver (and a higher diameter score as a result).

Distribution of Information Mapped to Network Density

Distribution of information can be mapped to the social network metric 'density', which is given by the formula:

$$Network\ Density = \frac{2l}{n(n-1)}$$

where *l* represents the number of links in the social network and *n* is the number of agents. The value of network density ranges from 0 (no agents connected to any other agents) to 1 (every agent connected to every other agent; Kakimoto et al., 2006). It is hypothesised that a peer-to-peer organisation will be denser than a hierarchical one, meaning that (all things being equal) broader dissemination of information will be rendered possible because there are more direct pathways between sender and receiver compared to a hierarchically organised counterpart.

Correcting for Scale

The need arises to check the behaviour of the social network metrics as a function of scale, and if necessary, to refine them further. In order to provide these checks we refer to two archetypal networks, the 'classic C2' (which in this case is constructed as a progressively deeper hierarchy as shown in Figure 5.2) and the 'fully connected peer-to-peer organisation' (which is the theoretical maximum represented by the so-called 'Edge Organisation', and also shown in Figure 5.2; Alberts and Hayes, 2005). These organisations should occupy polar opposite regions of the approach space (NATO, 2007) and the question arises as to how organisational size affects this.

Both network types started with three nodes and were increased over two further iterations to 7 nodes, then 15. Both sets of networks changed their characteristics but not necessarily their relative positions within the NATO SAS-050 approach space, depending on scale. Figure 5.2 shows that the peer-to-peer network's patterns of interaction and distribution of information remained stable in the face of increasing size. Regardless of size, the network remained fixed in a peer-to-peer configuration with every node within one edge of another node. Distribution of information remained at a constant, maximal level with everyone in the network potentially able to know everything. While these two factors remained constant as the peer-to-peer or edge organisation increased in size, one factor did change: allocation of decision rights. As more nodes joined the network it progressively moved along the axes towards greater distribution. Each new node joining the

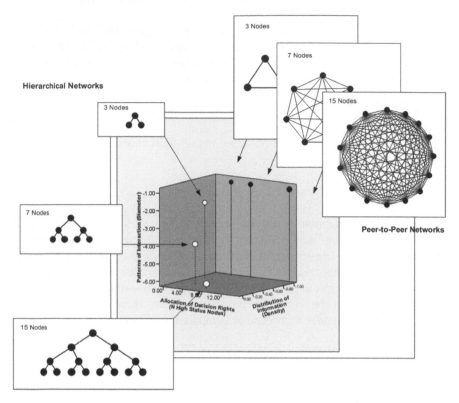

Figure 5.2   The effects of scale on the position that edge organisations and classic C2 occupy in the NATO SAS-050 approach space

network had the same level of interconnectivity, and the same potential role, as any previous node. This dimension behaves in a manner consistent with Leavitt's (1951) research presented in chapter three, in that it changes as 'a function of the size of the pattern as well as of its structure' (Leavitt, 1951, p. 47).

The behaviour of the classic C2 network is different. Figure 5.2 shows that as scale increased its characteristics varied over all three dimensions. Like the edge organisation, with increases in scale the distribution of authority became more distributed, but not to the same extent. Unlike the edge organisation the patterns of interaction became *more* hierarchical, with more layers being added to the network and more edges intervening as more nodes were added. In addition, the distribution of information shifted more towards the direction of 'tight control'. The classic C2 network became 'more archetypal' with increasing size whilst the peer-to-peer network was more or less as archetypal at large scales as it was at small scales. Classic C2, then, is more scale dependant in terms of its fundamental characteristics than the peer-to-peer 'edge organisation' which could be said to be relatively 'scale free'. Obviously, this is a very small analysis and constrained in certain ways because of it, but it does allow us to extract the general principle that scale does affect structure, in certain prescribed ways and more so for hierarchically disposed organisations than for edge organisations. Although only a very small analysis we are able to fall back on a much larger body of established work within organisational science with which these findings accord (e.g., Millhiser and Solow, 2007; Pondy, 1969; Carzo and Yanouzas, 1969).

Normalizing the Data

It is analytically expedient to not only correct for scale but to normalise the data so that each axis of the NATO SAS-050 approach space is represented by a simple percentage value. The corrections necessary to achieve this are as follows:

Normalizing allocation of decision rights Allocation of decision rights is defined numerically as the number of nodes who have a centrality value equal to or greater than the mean value. This measure is very scale dependant, as the number of nodes that can potentially meet this criterion increases with network size and thus could give a misleading insight into this dimension. In order to have different sized organisations sitting meaningfully on a common 100 per cent scale the number of nodes equal to or greater than the mean centrality value is expressed as a percentage of the total number of nodes in that particular network.

Normalizing patterns of interaction Patterns of interaction are ascribed the social network metric 'diameter'. The normalisation and correction is based on two simple network archetypes, Leavitt's (1951) chain network and Alberts' and Hayes' (2005) fully connected edge organisation. The chain network represents the maximum number of edges that have to be traversed in order to travel from one side of the network to the other ($d_{max} = n - 1$). The fully connected edge organisation represents the other extreme of 'minimum diameter' and the fewest number of edges from one side of the network to the other ($d_{min} = 1$). The following formula enables live measurements to be fitted between these two points and to be transformed into values ranging from 0 to 100 per cent:

$$\text{Patterns of Interaction} = \left(\frac{d-1}{n-2} \right) x100$$

where d = the obtained diameter measurement and n = the total number of nodes in the network. The formula applies across all values of n so that there is no scale related error. In the illustrations that follow, note that the result of this formula has been subtracted from zero in order merely to have the scale pointing in the right direction.

Normalizing distribution of information Distribution of information is ascribed the social network metric 'density'. The maximum value for density is 1, indicating that all nodes are connected to each other. Density is thus at base a simple proportion and can be multiplied up in order to have it expressed as a percentage, as shown below:

$$Network\ Density = \frac{2l}{n(n-1)}*100$$

Testing the Metrics Using Network Archetypes

The supposition that diameter, density and centrality can be used as metrics for decision rights, patterns of interaction and distribution of information can now be tested with reference to several theoretical network archetypes. Four of these are based on the early social network research by Bevelas (1948) and Leavitt (1951) who defined the 'Chain', the 'Y', the 'Circle' and the 'Star' networks as described in Chapter 3 and shown again in Figure 5.3. Plotting Leavitt's archetypes into the approach space enables regions within it to be anchored. Also, because these archetypes are accompanied by empirical data concerning their performance, this

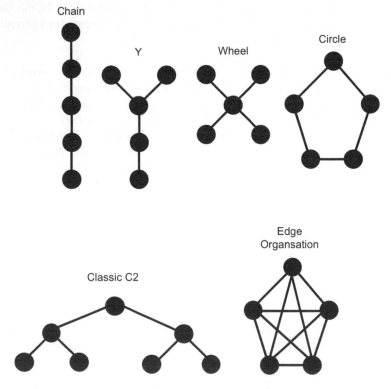

Figure 5.3   Illustration of archetypal networks. Associated with each is empirical evidence concerning its performance on simple and complex tasks

aids the goal of identifying crucial elements of a problem space and matching regions in this space to regions in the C2 approach. For example, the problem space might be suggestive of a task context that is complex, with the corresponding fix within the approach space being in close proximity to, say, the 'Star' archetype. On the basis of Leavitt's work it would be possible to not only make a crude judgement about this particular configuration being less than optimal but to outline more precisely why. In this case, networks exhibiting the properties of a 'Star' often overload the heavily connected high status node(s) in complex situations.

Bevelas and Leavitt's archetypes can be joined in the approach space by two further network structures derived explicitly from the theory behind the NATO SAS-050 approach space itself: the 'classic C2' organisation and the 'edge organisation' (also shown in Figure 5.3). The approach space proposes that these network archetypes should in theory fall into the bottom left and top right corners respectively. The hypothesis that diameter, density and sociometric status can be used as metrics for decision rights, patterns of interaction and distribution of information can thus be subject to a direct test: if the metrics work as expected, these two network archetypes should occupy similar positions in the approach space. Reference to Figure 5.4 shows this to be broadly the case.

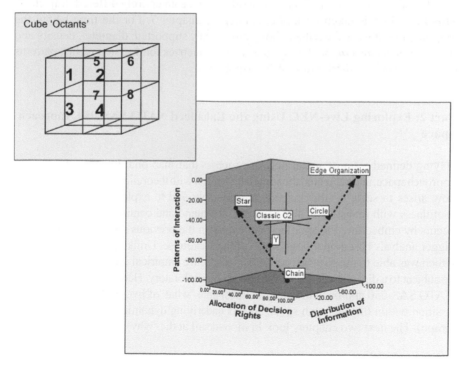

Figure 5.4   **Leavitt's network archetypes, along with the classic C2 and edge organisation archetypes, anchored into the enhanced NATO SAS-050 approach space**

Figure 5.4 shows that the classic C2 and edge organisation fall broadly into the areas of the approach space predicted. Although classic C2 is not pushed hard into the bottom/left/front position of the space as assumed in the NATO SAS-050 model, it is positioned in the correct 'octant'. As the discussion above regarding scale made clear, hierarchical networks tend to be much more scale dependent than comparable edge organisations, not to mention the fact that they can take many different forms from the simple archetype used in this analysis (e.g., deep, shallow, wide, narrow etc.). A point to note from Figure 5.2 is that more realistically sized hierarchies (of this simplified type) do indeed push further into the predicted part of the approach space. To the extent that such simple archetypes at least fall in the correct left/front/bottom 'octant' of the approach space means that in broad terms the model predictions are supported.

It is interesting to note that the circle archetype which in Chapter 3 we equated to an NEC organisation is not in actual fact a perfect match to the so-called edge organisation. However, it seems to sit on a distinct trajectory that springs from the chain network (the ultimate hierarchy?) upwards towards the edge organisation (the ultimate peer-to-peer?). In this regard the circle archetype is certainly quite distinct from the other archetypal networks like the Y, the Star and the classic C2 which also spring from the chain network but trace an entirely different trajectory. This behaviour is taken up and expanded in Chapter 7. For the time being, the mapping hypotheses described above are broadly supported: diameter, density and sociometric status *can* be deployed as useable metrics for decision rights, patterns of interaction and distribution of information.

Part 2: Exploring Live-NEC Using the Enhanced NATO SAS-050 Approach Space

Having defined a set of social network metrics that map onto the NATO SAS-050 approach space, and subjected those metrics to a test with theoretical data, an opportunity now arises to scale up the analysis considerably and to explore their efficacy and usefulness with realistically sized 'actual' command and control organisations. This occurs by embedding the approach developed in the previous section within a much bigger analysis based on a large scale military exercise. Unlike the previous section which was able to put forward a number of specific empirical questions which could be subject to a direct test, this section is entirely exploratory. Here we use the extended NATO SAS-050 approach to examine the simple 'what' of live-NEC in relation to its position within the approach space and its underlying dynamics (e.g., its agility and tempo). The next two chapters look in more detail at the 'why'.

Live-NEC

This analysis is based on a live case study. Data collection took place at a fully functioning brigade-level headquarters (BDE HQ) set up in a military training area for the purposes of evaluating a particular NCW system. The social network analysis did not focus on all forms of communication. It focused on 'inter-organizational' and 'inter-cellular' communications.

Inter-organisational communications took place between the BDE HQ and geographically dispersed battlegroup headquarters (BG HQ's). Several of these BG HQ's were simulated, along with enemy forces, from an experimental control centre (EXCON). This was geographically dispersed from both the live BDE and BG HQs. BDE HQ is in itself a reasonably sized organisation divided up into the conceptual equivalent of 'departments' (or cells). Inter-cellular communications refer to those that took place between different parts of the BDE HQ as these placed heavy reliance on the communications capability of the NCW system.

It is important to note that not all communications were able to be captured in this scenario. In particular, it was not practical or feasible to capture inter-personnel communications at a similar level of detail (although wider background to the more informal communications is found in Stanton et al., 2009). This has to be accepted as a limitation of the current analysis. Aside from that, the net-enabled command and control infrastructure was set up and staffed as it would have been if deployed. There was a total of 73 active data terminals in the scenario, 17 of which were located in and around the BDE HQ. There were also two encrypted-radio sets (and radio operators).

The operations phase took place over the course of a single day (with plans and so forth being prepared the day previously) and took 4 hours and 20 minutes to complete. In broad terms it comprised a rapidly approaching enemy from the west who had to be steered, through a combination of turn and block effects, to the northeast of the area of operations into a location where a 'destroy' effect would be deployed. Any remaining enemy units would then continue into the next area of operations which was not under the control of the present BDE HQ. It is important to note that the scenario was not a test of the military effectiveness of the military unit itself, and that the simulated enemy was also rather more compliant than perhaps is normally the case.

Data Sources

Two sources of data were used to inform the analysis. Firstly, comprehensive telemetry was extracted from the NEC system itself. The sampling rate of the telemetry varied but reached a maximum of approximately 10Hz, yielding a total of 2,866 data points pertaining to who was communicating to whom and it was from this that social networks could be created. This 'system log' data all resided at a 'digital' level in so far as it presented itself to the user through the NEC system's data terminals.

The second source of data was voice communications, which were transmitted over the encrypted-radio aspect of the NEC system. Data collection relied on a formal log of those communications kept by the incumbent of the watch keeper role. Every communication, its time, from whom it derived and to whom it was directed was recorded. This formed the basis of a social network analysis of inter-organisational 'voice' comms. Although mediated by a digital radio technology, the presenting modality of the communication, from the user's point of view, was 'voice'. A total of 158 discrete events of this type were extracted.

Modelling

Organisational Centre of Gravity

The 'digital' and 'voice' comms data was kept separate throughout the analysis as this provided an opportunity to see how the organisation varied across 'function'. To see how it varied over 'time' the data was divided into blocks of 85 comms events and entered into a proprietary software tool called WESTT (Workload, Error, Situation awareness, Tasks and Time; Houghton et al., 2006 and 2007). This tool enabled social networks to be created for each block of 85 comms events.

Over the course of the four hour and twenty minute exercise period the digital comms data amounted to 34 blocks of 85 comms events in total. The voice comms data was then spread across the same 34 blocks, and of course, being less numerous than the digital comms they did not fill the 85 discrete comms events per block that the digital comms achieved. Indeed, no voice comms events at all took place within the first and last data blocks, so in the case of voice comms just 32 social network analyses were produced.

Associated with each of the 34 social networks for digital comms, and 32 networks for voice comms, is an accompanying density, diameter and centrality metric. The mean of these values represents the set of coordinates which describe the organisations centre of gravity within the approach space. The individual values create the data necessary to analyse organisational agility as follows.

Agility   The organisation's movement around its centre of gravity was visualised by using each of the 32 (voice) or 34 (digital) social networks to plot the organisation into the approach space. By these means it becomes possible to see how the organisation's location varied across time (with each data block representing a time interval) as well as function (voice versus digital communications). The range of values that density, diameter and centrality describe over time, also describe the organisation's ability to change and reconfigure itself. According to Alberts and Hayes (2006) this provides a measure of agility.

Tempo   When the values for density, diameter and centrality are each plotted onto a graph with the *x*-axis representing time and the *y*-axis representing the corresponding social network metric value, the resultant line resembles a waveform. Spectral analysis methods can then be used in order to say something about 'tempo'. Spectral analysis methods, broadly speaking, decompose time-varying data like this into frequency components using Fourier Analysis. The principle here is that any complex waveform can be separated into its component sine (or pure) waves and the resultant data plotted into a graph called a periodogram. A periodogram has a frequency scale on the *x*-axis measured in Hertz (or cycles per second) so that moving right along the *x*-axis means that density, diameter or centrality fluctuate more quickly (i.e., higher tempo). As well as speed of fluctuation there is also the strength of that fluctuation, which is ascribed a power value on the *y*-axis. The higher the power value the stronger the fluctuation at that particular frequency. Such an analysis, therefore, provides a surprisingly literal representation of 'battle rhythm', albeit subject to a modifying caveat. The measurement intervals used in the current analysis distort the frequency scale somewhat (the sampling rate is not in cycles per second). This means that it is not safe to interpret the *x*-axes of the following periodograms literally but rather to interpret them 'relatively' as 'order of magnitude effects'.

The periodogram represents a fascinating diagnostic tool in terms of it being able to detect the presence of meaningful temporal patterns in the way that the organisation configures and reconfigures itself under real-life dynamic conditions. If meaningful patterns exist then something of their character can be revealed: evidence of periodicity indicates that the organisation is repeatedly drawn into specific locations of the approach space. Just as the presence of periodicity communicates something interesting so to does the lack of a pattern or periodicity. It suggests that the forces present in the problem space which are driving the organisation's dynamic reconfigurations are more chaotic in nature.

Results: Digitally Mediated Communications

Centre of Gravity

The organisational centre of gravity for the digital comms function is specified by the following three coordinates:

1. Mean decision rights (centrality) = 7.44 per cent
2. Mean patterns of interaction (diameter) = -12.78 per cent
3. Mean distribution of information (density) = -78.17 per cent

This gives rise to the digital comms function occupying the position shown by Figure 5.5. This is a function characterised by unitary decision rights but fully distributed patterns of interaction and broad dissemination of information. On face value, then, a strange hybrid that is neither classic C2 nor a fully connected edge organisation, nor indeed any of the other more intermediate forms of organisation which lie in between.

Figure 5.5 shows that live-NEC does not fall into any regions that the various network archetypes plot into. The closest archetypes appear to be the classic C2 and star networks. As we saw in Chapter 3, these networks tend

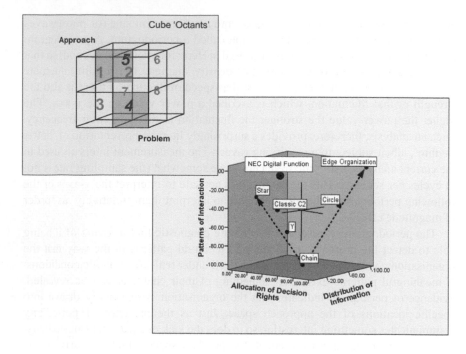

Figure 5.5   Organisational centre of gravity for live-NEC case study (digital comms function). Live-NEC (the approach) occupies cube octant #5 whereas the extant problem (i.e., a cold-war style engagement) occupies octant #4

to be associated with people funnelling information to the centre upon which answers are sent out. But given that both these archetypes fall into a different cube octant to that of live-NEC this would seem to be only a weak attribute of this particular region of the approach space. The fact that a live organisation has plotted into this region does suggest, however, that there may be something interesting about it.

The next observation concerns the mapping to the corresponding area of the problem space. The NATO SAS-050 problem space currently lacks the metrics to make a precise problem definition but even at a crude level of analysis there is evidently a mismatch. The scenario in this case was an overtly cold-war style of engagement characterised by high familiarity (of enemy doctrine), fairly static rates of change (to the extent that the dynamics are more or less linear and in sequence) and a strong information position (a lot is known about enemy capability and position). The corresponding octant in the problem space in this instance would be the bottom/*left*/*front* rather than *top*/left/*back* octant. The fact that there appears to be a mismatch is born out by the sometimes extreme difficulty that staff had in cajoling the system to meet their needs (again, the full extent of this is described at length in Stanton et al., 2009).

Tempo

If the communications network was stable then the representation shown by the centre of gravity described above would be sufficient. It is not. The organisation is far from stable because the humans in it reconfigured it in response to their environment/problem. This reconfiguration is clearly evident when all 34 sequential social networks from the digital comms function are plotted into the approach space along with the archetypal networks. What results is a form of 3D scatter plot which clearly illustrates the dynamical behaviour of the digital function over time.

Figure 5.6 shows that the values of x (decision rights/centrality), y (patterns of interaction/diameter) and z (distribution of information/density) are time varying. They thus become amenable to alternative analysis techniques which help to shed light not just on the different regions of the approach space which can be occupied

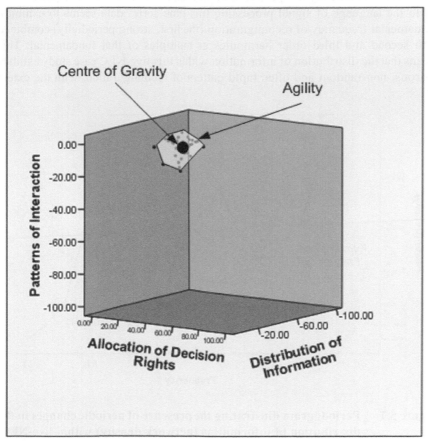

Figure 5.6 **Results of plotting all 34 digital comms networks into the enhanced NATO SAS-050 approach space**

(agility), but the speed and periodicity with which different reconfigurations are undertaken.

Figure 5.7 is a periodogram of network density. This is the output of the spectral analysis method described earlier and is designed to detect the presence and strength of any periodicity in the data. A considerable amount of analogy and conjecture is at work here and strict scientific rigour has been sacrificed for innovation and novel insights into tempo. To that extent the *x*-axis (marked frequency) represents a scale with very slow fluctuations in the distribution of information (density) at the leftmost end, to rather more rapid fluctuations at the other. The pattern of results gained is intriguing. They show a power spectrum dominated by very low speed fluctuations in distribution of information accompanied by the presence of what appear to be second and third order harmonics of diminishing strength but increasing frequency. In other words, overlain on this slowly cycling level of information distribution is a higher frequency of network reconfiguration but at reduced power. Overlain on top of this is yet another higher speed reconfiguration even further reduced power.

In the language of signal processing this time series data seems to exhibit a fundamental frequency of reconfiguration (the first, strong periodicity) combined with second and third order harmonics at multiples of that fundamental. This means that the distribution of information within this live-NEC case study exhibits a strong, non-random and often rapid pattern of reconfiguration, with the extent

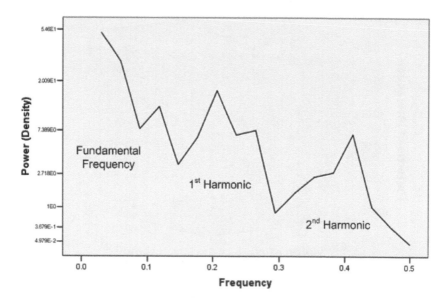

Figure 5.7   Periodogram illustrating the presence of periodic changes in the distribution of information (network density) within live-NEC. A pattern is obtained that approximates to first, second and third order harmonic effects

to which people have access to relevant information varying quite widely at the 'broad dissemination' end of the spectrum. More fundamentally, the underlying theory governing this dynamical behaviour, because it has strong periodicities (and therefore regularly repeats itself) appears to be deterministic in character.

Patterns of interaction (given by the metric 'diameter') also exhibit interesting behaviour (Figure 5.8). Again, there is the presence of what appears to be a strong fundamental frequency suggesting an underlying rate of expansion and contraction in the extent to which the organisation is a fully distributed type. There is also a feature of Figure 5.8 that could be taken for something vaguely analogous to a second harmonic occurring at a multiple of the fundamental periodicity. What this graph communicates is that there are strong and non-random changes in the patterns of interaction, with comparatively rapid shifts occurring between fully hierarchical and fully distributed modes of operation. Again, high tempo seems to be in evidence.

The pattern of results observed above for the distribution of information (density) and patterns of interaction (diameter) are not similarly evident for allocation of decision rights (centrality). The periodogram (not reproduced in this case) exhibits a more or less random pattern, a lower, more uniform power spectrum and little evidence of systematic periodicity. In these terms, there is evidence that movement along the scale from unitary to peer-to-peer decision rights is rather low in tempo (relatively speaking) and governed by more non-deterministic factors. In other words, the fact that this organisation was in one state of 'decision rights' is not a very good predictor of the next or future states.

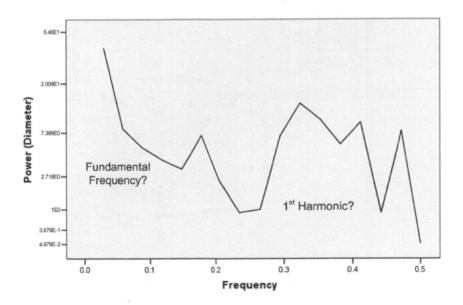

Figure 5.8   **Spectral analysis graph illustrating the presence of periodic changes in the pattern of interaction (diameter)**

Results: Voice Mediated Communications

The analysis performed on the digital comms function can now be repeated for voice comms. As a lot of the explanatory ground work has already been covered above, this section can be considerably briefer and to the point.

Centre of Gravity

The organisational centre of gravity for the voice comms function is specified by the following three coordinates:

1. Mean decision rights (centrality) = 37.34 per cent
2. Mean patterns of interaction (diameter) = -11.35 per cent
3. Mean distribution of information (density) = -25.62 per cent

This gives rise to the voice comms function occupying the position shown in Figure 5.9. This is a function characterised by somewhat less unitary decision rights (centrality) than the digital comms function. The voice function has comparable

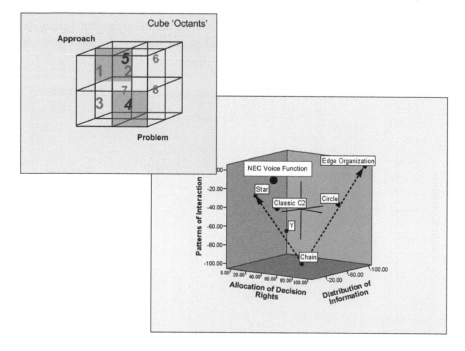

Figure 5.9   **Organisational centre of gravity for the live-NEC case study (voice comms function). Live-NEC (the approach) occupies cube octant #5 whereas the extant problem (i.e., a cold-war style engagement) occupies octant #4**

patterns of interaction (diameter) but a somewhat lower distribution of information (density). In the case of voice comms, then, despite some differences between the comparable digital comms function they both still fall into octant #5 of the approach space. Although the voice comms function is pushing slightly nearer to octant #1 than the comparable digital function, it too is a strange hybrid that is neither classic C2 nor a fully connected edge organisation. The same point about the approach matching the extant nature of the problem holds in this section also.

Tempo

The same analysis of tempo described above can be repeated here. This time 32 social networks pertaining to the reconfiguration of the voice comms networks are plotted into the enhanced approach space to yield Figure 5.10.

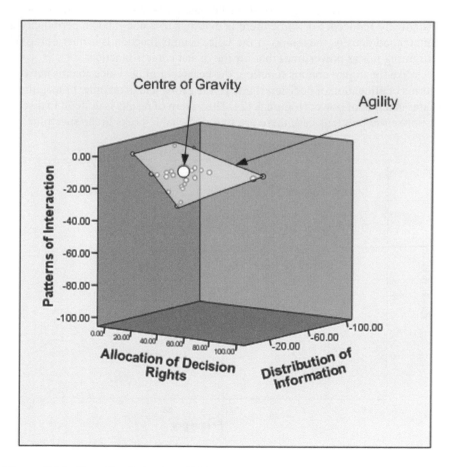

Figure 5.10   Results of plotting all 32 voice comms networks into the enhanced NATO SAS-050 approach space

Reference to Figure 5.6 and Figure 5.10 shows, visually at least, that the voice and digital comms functions differ in terms of their underlying dynamics. Spectral analysis may help to shed more light on this.

Figure 5.11 is a periodogram of network density (i.e., distribution of information) for the voice comms function. The pattern of results is more or less identical to those gained for network diameter (i.e., allocation of decision rights) and given the similarity and 'order of magnitude' interpretation being placed on these periodograms only that for density need be reproduced. Figure 5.11 shows a power spectrum dominated by a very low speed fluctuation in network density and diameter. It communicates the fact that the network configures and reconfigures itself, and is able to alternate between 'hierarchical' and 'fully distributed', and from 'tight' to 'broad dissemination'. Compared to the digital comms function, however, there is a relative absence of high-frequency harmonics (or other discernable components) and it is of further interest to note that the power level is substantially reduced. So, while there is evidence of a non-random periodicity in diameter and density, the tempo of the voice comms function is neither as rapidly configuring nor as powerful as that for the digital comms function.

Unlike the digital comms function, the behaviour of the voice comms network in terms of allocation of decision rights is interesting and of an order of magnitude greater in terms of power (Figure 5.12). The pattern of results is difficult to discern but unlike the previous case there are two discernable peaks in the spectrum that

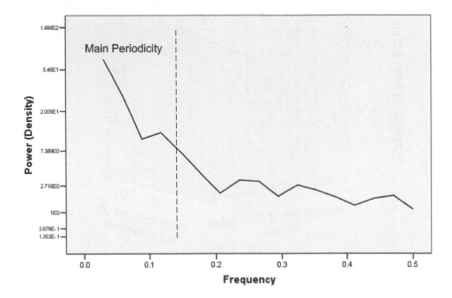

Figure 5.11   Spectral analysis graph illustrating the presence of periodic changes in network density. A more or less identical pattern of results is achieved for network diameter

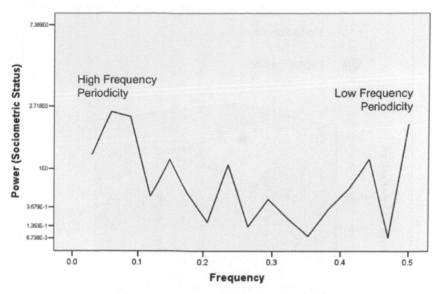

Figure 5.12   Spectral analysis graph illustrating the presence of periodic changes in high-status nodes

indicate a low and high speed change in the number of high status nodes. On this measure, then, the voice comms function appears to be governed by more deterministic forces than the comparable digital function (which exhibited greater degrees of chaotic behaviour). In summary, both voice and digital functions were able to demonstrate an ability to reconfigure rapidly and exhibit what could be regarded as high tempo, the digital function somewhat more so.

Results: Agility

Figure 5.13 provides a simple visual diagnostic for the total area in the approach space that is bounded by the positions that both the digital and voice comms functions occupy within the approach space. Compared to the sophisticated visualisation techniques found in the physical sciences (for example, convex hulls, manifolds and splines) merely linking the points that form the perimeter of the scatterplot (as in Figure 5.13) is perhaps overly simplistic. As a first pass, however, it serves to illustrate Alberts and Hayes (2006) expedient that the number of distinct points in the approach space are what defines an organisation's agility. Furthermore, the levels of agility achieved by the different functions within the live-NEC case study (digital and voice) appear at first glance to be different. This warrants further exploration.

The data in Table 5.1 provides a more formal analysis of the total area in the approach space occupied by the different comms functions. It presents the

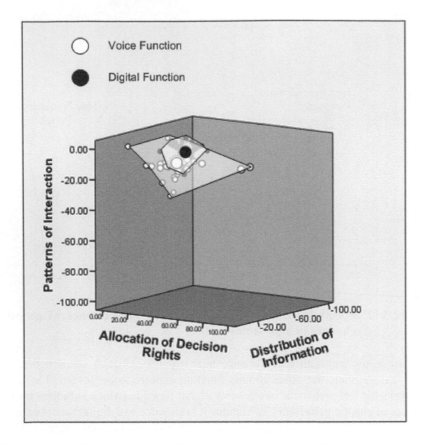

Figure 5.13 **Visual representation of the total region(s) occupied by the digital and voice functions contained within the live-NEC case study**

Table 5.1 **The range of values on each of the enhanced NATO SAS-050 model axes and a composite measure of agility (all data shown as %)**

	Decision rights (x)	Patterns of interaction (y)	Distribution of information (z)	Agility*
Digital comms	13.13	21.30	39	24.48
Voice comms	100	34.38	80	71.46
Total system	100	34.38	85	73.13

* Total figure based on summing x, y and z figures, dividing by the total maximum per cent available in x, y and z columns (i.e., $3 \times 100 = 300\%$) and multiplying by 100.

difference between the minimum and maximum values on each of the model dimensions and a simple aggregate measure of agility based on the total of these figures. Figure 5.14 provides a graphical summary of the same.

Figure 5.14 is a composite representation of the amount by which the organisation varied along its three measured social network metrics across digital and voice functions. This highly simplistic measure of change represents the range of coordinates in the 3D space and says something about the number of locations in that space which were actually occupied. These three values were summed, divided by the maximum percentage value achievable in each x, y and z column (i.e., 3×100 per cent) then multiplied by 100 in order to provide a simple percentage score. Note that the x, y and z values are based on a set of calculations that normalise the data and offer some correction for network size. This, then, is a very crude measure of agility but to the extent that it offers some kind of comparison it certainly supports the visual impression shown in Figure 5.13, where it can be seen that the networks associated with the voice comms function are more widely scattered.

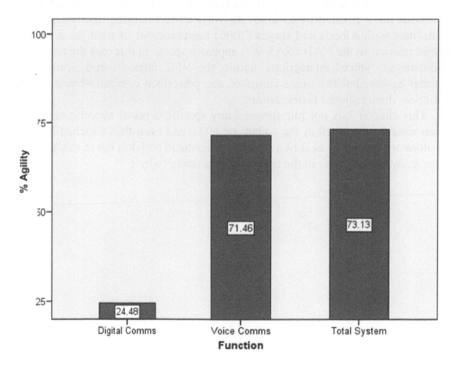

Figure 5.14   A rough order of magnitude measure of total agility shows that the voice-based communications architecture in the live-NEC case study was more agile than the digital architecture

Summary

This chapter has advanced the sociotechnical analysis of NEC to the point where we are now tentatively dealing with a live case study. The key innovation has been to use the social network analysis techniques first mentioned in Chapter 3 in order to define quantitative metrics for each of the NATO SAS-050 approach space's primary axes (alluded to in Chapter 1). It is this simple expedient that has provided a practical means to define the actual place or region in the approach space where this 'real-life' organisation operates, to see how that ranges over function (digital and voice communications) and time (by taking numerous slices through the data). Through the use of well-understood network archetypes, regions within the approach space have been anchored to empirical data which in turn has helped to facilitate a comparison between the approach and the problem to which it is directed. In this regard we note a mismatch.

Social network analysis lies at the heart of this chapter and the method further differentiates itself by being a form of dynamic social network analysis (e.g., Monge and Contractor, 2003). The changing values of the network based metrics are what have enabled us to innovate ways of representing agility and tempo consistent with Alberts and Hayes (2006) interpretations of what these concepts are in relation to the NATO SAS-050 approach space. In this case the naturalistic radio-based 'voice' interactions within the NEC infrastructure demonstrated greater agility, but the more complex and prescribed computer-based 'digital' function, demonstrated faster tempo.

This chapter has not put forward any specific a priori hypotheses and has been aimed very much at the 'what' of NEC, and even then couched at a very exploratory level. The next two chapters continue to pull data out of this live-NEC case study but this time in the course of examining 'why'.

Chapter 6

The Human in Complex Dynamic Systems

Aims of the Chapter

In the previous chapter we used the enhanced NATO SAS-050 approach space to understand something of the 'what' of live-NEC. In this chapter we revisit the topic of complexity to understand a little of the 'why'. The core concepts of emergence, open systems behaviour, sensitive dependence on initial conditions and dynamical system behaviour are dealt with from a human-factors point of view, and with reference to the findings already extracted from Chapters 3 and 5. Emergence expands on the human-factors issues surrounding user adaptability, sensitive dependence relates to the inherent instability of the human/system interaction, open-systems behaviour considers the data/information dichotomy, and dynamical-system behaviour deals with how complex systems can be understood and modelled. In each case, a concrete example of how complex systems research can enhance the study of human factors is presented.

Introduction

In Chapter 4 complexity was distilled through three overlapping views. There was the attribute view, in which complexity is about the multidimensional space in which human factors problems are located. It was conjectured how this multi-dimensional space could be collapsed into three critical variables in order to create a defined 'problem space'.

Then there was the complex theoretic view, which deals with the quantification of certain aspects of complexity. Here it was shown how the metrics provided by this approach complement existing mainstays in human factors methodology, allowing better judgements to be made about whether one entity or phenomenon is numerically more complex than another.

Finally there was the complex-systems research view, the domain of 'true complexity', of the emergent phenomena occurring at the edge of chaos and the phenomenon which require a new set of tools and approaches in order to diagnose what is going on and what to do about it. Here, complexity was related to the 'approach' needed to match putative 'problems'. In other words, complex problems (i.e., those that are unfamiliar/unstable/unknowable) require systems to be configured in certain ways (i.e., peer-to-peer interaction, devolved decision rights and widespread dissemination of information). Complex problems also seem to require human factors itself to be configured in certain ways, for its various 'approaches' to match its various 'problems'. In this chapter it is time to drill down further in order to extract even more specific insights and methods. Theory is turned into practice by relating the findings from Chapters 3 (Leavitt's communications patterns study) and 5 (the live-NEC study) to the study of complexity, and relating the sum total of that back to human factors.

Emergence

Definitions

In some senses, complexity *is* the science of emergence (Waldrop, 1992). In colloquial terms, emergence is the 'more than the sum of its parts' effect of synergy, of 'getting something for nothing' or even 'where something stupid buys you something smart'. Emergence describes behaviour that is not deducible from its low level properties. At one end of the spectrum are emergent properties that appear as mysterious and deeply perplexing, at the other are a vast array of emergent phenomena which are entirely mundane and prosaic. Negotiating a route between these two extremes requires a workable definition of the concept that relates to human factors problems:

> 'Emergence is the phenomenon wherein *complex, interesting high-level function* is produced as a result of combining *simple low-level mechanisms in simple ways.'* (Chalmers, 1990, p. 2)

Types of Emergence

Table 6.1 opposite shows that not all emergence is the same.

Diagnosing Emergence

Table 6.1 shows that: 'As systems become more complex [...], self-organisation appears at more than one level [...]. Such systems have multiple, hierarchical levels of self-organisation, and calculation of system level emergent properties from the component level rapidly becomes intractable' (Halley and Winkler, 2008, p. 12). What this simply means is that in some cases it can be easier to describe and predict the emergent behaviour itself rather than its detailed component level antecedents. The question, then, is how to tell between the levels and decide what level of emergence one is confronted with? Solutions to this question are available for certain kinds of physical problem but the situation is considerably more difficult for human factors problems. One approach that offers certain useful analogies is the concept of Relative Predictive Efficiency (RPE; Crutchfield, 1994). RPE can be expressed as follows:

$$RPE = E/C$$

Table 6.1 **Types of emergence, the information needed in order to make a diagnosis of emergent phenomena and the associated level of difficulty in doing so (definitions from Bar-Yam, 2004c)**

Emergence	Type	Information needed in order to make a diagnosis of collective system behaviour	Difficulty in diagnosing collective system behaviour
None	Deterministic	Knowledge of individual system components sufficient to fully explain global system behaviour.	(Relatively) easy…
Weak	Type 0*		
	Type 1	As for Type 0 but with additional knowledge about the positions and dynamics of individual entities in a system, this being sufficient to describe the 'microscopic as well as macroscopic properties of the system' (Bar-Yam, 2004c, p. 17).	…more difficult…
Strong	Type 2	As for Type 1 but with additional knowledge of possible states and configurations the system can adopt. 'the state of one part may determine (or be coupled to) the state of other parts' (Bar-Yam, 2004c, p. 17).	…extremely difficult
	Type 3	As for Type 2 but with additional knowledge of the environment that the system resides in. 'This is not contained in the conventional discussion of properties of a system as determined by the system itself' (Bar-Yam, 2004c, p. 17).	
Random	Chaotic	Complete and absolute knowledge of all entities, interactions and possible interactions in the future.	…Impossible (for most practical purposes)

* This form of emergence carries with it a variety of epistemological issues which are beyond the scope of this book. Suffice to say that at least some forms of emergence can be understood from knowledge of parts alone.

E is 'excess entropy', or for practical human factors purposes, the extent to which a system can be adequately modelled. In fields that deal with physical systems the extent to which (typically mathematical) models are able to predict a given behaviour is often relatively straightforward. It is rather less so in the human sciences. At the crude and overly simplistic level there is a comparison to be made between the system behaviours predicted by a model (say, HTA) compared to those behaviours actually observed. Any disparity between 'expected' and 'observed', and in what quantity, could represent some equivalent of 'excess entropy', or 'E' in the formula. C, on the other hand, is 'statistical complexity'. It is a measure of the size and/or complexity of the system's model at any given scale of observation. In this respect the metrics presented in Chapter 4 under Complexity Theory seem well suited.

Imagining for a moment that suitable metrics for E and C can be derived and/or approximated, RPE enables the analyst to decide what approach to take: strict reductionism (and a focus on component antecedents of system behaviour) or systemic (and a focus on the system's emergent behaviour itself). Emergence exists, and therefore systemic methods become more appropriate, 'if the higher level description of the system has a higher predictive efficiency than the lower level' (Halley and Winkler, 2008, p. 13). This links back to the discussion in Chapter 4 about complexity and scale and once again we see an interplay between the scale of observation and the scale of behaviour (e.g., Bar-Yam, 2004a). In this case RPE provides insight into the type of analysis suited to a particular complex system, and in this respect could lead to considerable savings in analysis time and effort for a corresponding increase in predictive efficiency. We can examine this in a little more detail by referring back to the live-NEC case study described in Chapter 5.

Emergence in Live-NEC

Communications content   In the last chapter we saw how 2866 digitally mediated comms events took place during a live-NEC field trial. As well as measuring the presence and frequency of these communications we also analysed their type. The digital comms function provided a comms 'type' label within the raw 'system logs' that generate the data and these were preserved and subsequently interpreted in the analysis. The different types of communications extant on the digital function of the NEC system are explained in Table 6.2.

The emergence of 'Free Text'   Figure 6.1 on page 108 shows the content of what is being transmitted on the digital comms layer.

The largest proportion of transmissions is accounted for by System Messages. These relate to the NEC system's status and current activity (37 per cent). This is followed by acknowledgements that something has been received, or of an action that has been performed on the system (30 per cent). Next in line, and perhaps most interesting in the present context, is Free Text (9 per cent).

Free Text, as the name implies, is the facility for users to type messages without the imposition of any kind of predetermined layout, format or proforma. In other words, it is a catch-all form of functionality which, given the highly structured interaction provided by the system, is not one that the designers seemed to have anticipated being used all that extensively. In practice, however, it seems clear that personnel started to rely on it to a far greater degree than expected. They combined this simple, low-level mechanism to create interesting high-level emergent function.

Table 6.2   **Definition of digital communication types that occur during live-NEC case study**

Comms type	Description
Acknowledgements	Acknowledgements that a communication has been sent/received or some other action has been performed on the system.
System Messages	Warnings, advisory notices and information automatically generated by the NEC system.
Free Text	A type of communication not constrained to pre-defined templates or proformas (a crude equivalent of email).
CASREP	Casualty report.
COMBATREP	Combat report.
SIGHTREP	Sighting report.
INTOLAY	Intelligence overlay.
OWNSITREP/situation	Own situation report.
PERSREP	Personnel report.
NBCSIGHTREP	Nuclear, biological, chemical sighting report.
OpO	Operations order.
Position	Non-automated position report.
Overlays	Digital equivalents of map overlays.
WngO	Warning order.
IntReps	Intelligence report.
Military messages	Information from higher command centres.
Own Position	Automatically generated GPS coordinates to enable live tracking of assets.

Binomial tests demonstrate a persuasive non-random effect when Free Text is compared in a pairwise fashion to various other forms of 'constrained text' (i.e., communications in the digital domain that are constrained by templates, pre-defined formats and so on). Table 6.3 presents the results of this comparison.

Table 6.3 shows that the proportion of Free Text comms is statistically greater than any other individual form of constrained communication; therefore, for an 'additional feature', one whose use is not actively encouraged in standard operating procedures, it seems to have assumed a much more important role than anticipated.

Moving from digital comms transmitted (n = 454) to those that were received (n = 2412) a similar pattern of results is obtained. As Figure 6.2 shows, by far the largest proportion of incoming digital comms is accounted for by positional data

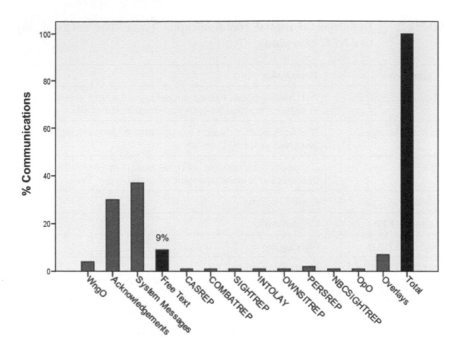

Figure 6.1   Bar chart showing the type of data comms that are being transmitted by BDE HQ

Table 6.3   Free versus 'constrained' digital comms that are 'transmitted'

	N	Observed proportions		
Free Text	39	Constrained	Free Text	Exact Sig (2-tailed)
CASREP	4	0.09	0.91	p<0.0001
COMBATREP	3	0.07	0.93	p<0.0001
SIGHTREP	2	0.05	0.95	p<0.0001
INTOLAY	2	0.05	0.95	p<0.0001
OWNSITREP	4	0.09	0.91	p<0.0001
PERSREP	11	0.22	0.78	p<0.0001
NBCSIGHTREP	1	0.03	0.98	p<0.0001
OpO	1	0.03	0.98	p<0.0001

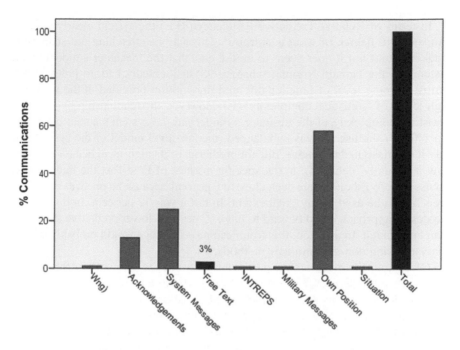

Figure 6.2   Bar chart showing the type of data comms that are received by BDE HQ

broadcast by sub-units (58 per cent). Note that the system uses this automated data to display icons on a map display. This is followed by system messages (25 per cent) and acknowledgements (13 per cent). It is interesting to note again the relative prominence of 'Free Text'. Although it only accounts for 3 per cent of total comms it is fourth highest. A similar finding to the above (i.e., Table 6.3) is detected in that Free Text accounts for a greater proportion of user initiated comms than the other, more constrained types, as Table 6.4 shows.

Table 6.4   Free versus 'constrained' data comms that are 'received'

	N	Observed proportions		
Free Text	65	Constrained	Free Text	Exact Sig (2-tailed)
INTREPS	4	0.03	0.97	p<0.0001
Military messages	3	0.06	0.94	p<0.0001
WngO	2	0.16	0.84	p<0.0001

In terms of Relative Predictive Efficiency (RPE) the system clearly exhibits a meaningful degree of 'excess entropy'. This may be stretching the analogy to breaking point but it does seem to be the case that the 'designer's model of the system' (in the Donald Norman sense; 1998) undoubtedly had, to judge by the almost extreme level of functionality and prescription provided in the interface, high levels of 'statistical complexity' (C). However, the relationship between the statistical complexity of the designer's model and the system's actual behaviour (specifically the user's) was unbalanced: the low-level model of the system, the way it was designed to behave, did not predict all of its high-level behaviours (i.e., how it 'actually' behaved). In the specific instance of Free Text the figure E (for excess entropy) should have been close to 0 per cent because no one was expecting Free Text to be used to any great extent. In fact it was 12 per cent. In this respect the design approach could be said to have 12 per cent lower predictive efficiency than anticipated. In addition, this troublesome percentage would probably best be analysed using non-deterministic methods.

Of course, there is nothing particularly new in the observation that people use systems in unexpected ways (e.g., Lee, 2001; Roth et al., 2006). This merely reflects a long-standing sociotechnical principle of design, that the human system interaction is not stable and that the people using it should be expected to 'interpret it, amend it, massage it and make such adjustments as they see fit and/or are able to undertake' (Clegg, 2000, p. 467). From a complex systems research point of view, excess entropy seems to represent individuals increasing the system's variety at finer scales and thus increasing the system's ability to cope with complexity. From a human-factors point of view the function that phenomena like this serves is as an indication of the type of interaction that users are trying to design for themselves. In the case of Free Text the users were faced with complexity and they decided to use a simple type of interaction (Sitter, Hertog and Dankbaar, 1997; Rothrock et al., 2002). An approach to design and analysis which acts in sympathy with this phenomenon is likely to yield higher predictive efficiency so that it becomes a case of simpler (systemic) models with greater predictive efficiency, as opposed to complex (reductionist) models with less. We can explore this further by looking at the more naturalistic communications afforded by the voice functionality of the live-NEC system.

Simple interactions for complex tasks   Whereas the digital function of the live-NEC system is an example of complex interaction, the voice comms function using the radio is comparatively simple. If users favour a simple interaction that allows them to do more complex tasks then this should be reflected in the type of communications that exist on the voice function of the live-NEC system. To examine this a little more closely, seven communications typologies are defined with every instance of an inter-organisational verbal exchange (n = 158) being categorised as shown in Table 6.5.

The results of this categorisation are shown in Figure 6.3 for both transmit (n = 45) and receive (n = 113) events.

Table 6.5   Bowers et al. (1998) communications typology

Factual	'objective statement involving verbalized readily observable realities of the environment, representing "ground truth".'
Meta-query	'request to repeat or confirm previous communication'.
Response	'statement conveying more than one bit of information' (i.e., comprising of more than simply 'yes/no').
Query	'direct or indirect task-related question'.
Action	'statement requiring team member to perform a specific action'.
Acknowledgement	'one bit statement following another statement (e.g., "yes", "no")'.
Judgement	'sharing of information based on subjective interpretation of the situation' (Cuevas et al., 2006, p. 3–4; Bowers et al., 1998).

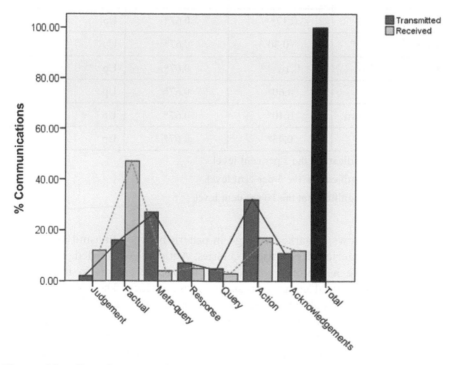

Figure 6.3   Bar chart showing the content of voice comms transmitted and received, according to Bowers et al.'s (1998) taxonomy

Using the multiple regression based technique explained in Chapter 3 (Warm, Dember and Hancock, 1996) each individual communications category can be analysed in terms of its relative contribution to the aggregate 'total communication' score. The contribution is expressed as a standardised beta coefficient, the higher the beta coefficient, the more that a comms subscale contributes to total communications. This method provides an order of magnitude analysis highly useful for discerning patterns of communication such as this the results of which are shown in Table 6.6.

Table 6.6 **Standardised beta coefficients show the relative contribution that each voice comms subscale makes towards total communications (for both transmit and receive events)**

	Beta coefficients		
	Receive	**Transmit**	**Direction of change**
Factual	0.73*	0.33*	Down
Meta-query	0.17**	0.67*	Up
Response	-0.30	0.67*	Up
Query	0.16***	0.67*	Up
Action	0.60*	0.67*	Up
Acknowledgement	0.40*	0.67*	Up
Judgement	0.43*	0.67*	Up

* Statistically significant at the 1 per cent level

** Statistically significant at the 5 per cent level

*** Statistically significant at the 10 per cent level

Table 6.6 shows a marked change in pattern between transmit and receive events. Meta-Queries, Responses, Queries, Actions, Acknowledgements and Judgments make a uniform contribution to overall comms in the transmit category, with Factual comms assuming a lesser place. The output, in voice comms terms, is less about 'objective statements involving verbalized readily observable realities of the environment' and rather more to do with 'requests to repeat or confirm previous communication, statements conveying more than one bit of information (i.e., comprising of more than simply 'yes/no'), direct or indirect task-related questions, statements requiring team members to perform a specific action, one-bit statements following another statement (e.g., 'yes', 'no') and the sharing of information based on subjective interpretations of the situation'. In this regard voice comms emerges as rather similar to the sorts of interactions being undertaken using Free Text, a form of recursive, redundant, informal interaction driven by circumstances and not just templates, proformas and other prescribed ways of working. This pattern switches for comms that are received where the focus is very much on observable realities of the situation. The picture that emerges from this dichotomy is one of 'data' versus 'information'.

Open Systems Behaviour

With two separate encoding taxonomies existing between digital and voice functions, a way to compare across them is required. For this a cautious distinction between 'data' and 'information' is made. Caution is exercised because defining important terms like these, theoretically at least, remains surprisingly contentious. As a result, nothing more than a crude attempt to impose this categorisation is attempted and even then it is intended that 'data' and 'information' serve merely as broad descriptive labels. A tolerable stab at defining these terms is made by anchoring them to Endsley's three-stage model of SA (1988).

Endsley (1997) points out that success in something like the live-NEC case study's endeavours 'involves far more than having a lot of data. It requires that the data be transformed into the required information in a timely manner'; furthermore, 'Creating information from data is complicated by the fact that, like beauty, what is truly "information" is largely in the eyes of the beholder' (p. 2). If Level 1 SA refers to 'the perception of elements in the environment', and for current purposes emphasis is placed on 'elements in the environment', then these we might refer to simply as 'data', that is, as objective, measurable realities of a situation. Level 2 SA refers to comprehension of what those elements might mean (thus, in the eye of the beholder, mere data is becoming information) whilst Level 3 SA refers to 'projecting their status into the near future' (in other words, actually doing something with the information). Grouped together, Level 2 and 3 SA seem to bound a category that is undoubtedly more 'information-like' than it is 'data-like' (or Level 1 SA-like).

Table 6.7 shows what the different comms layers (digital and voice) predominantly carry; data or information?

Table 6.7   Data vs. information received and transmitted by Brigade Headquarters

	Digital %	Voice %
Level 1 ('data')	73	47
Levels 2 and 3 ('information')	27	53

From Table 6.7 it is clear that the digital comms layer carries a far higher proportion of 'data' compared to the voice layer, which carries a greater proportion of 'information'. This disparity is statistically significant ($\chi^2 = 14.08$; df = 1; p < 0.0001, and is accompanied by a moderate effect size; Cramer's V = 0.27; p < 0.0001). Note that the Chi Square test has been performed on the percentage of the raw scores. The reason for this is that statistical power becomes excessively high with 3,000 or so data points thus making it extremely unlikely that a statistical difference would not be detected. Table 6.8 contributes further to this picture. It goes on to show that it is predominantly 'data' that is received by the headquarters (96 per cent) compared to a greater proportion of 'information' leaving it (26 per cent).

Table 6.8   Data vs. information received and transmitted by Brigade Headquarters

	Received by BDE	Transmitted by BDE
Level 1 ('data')	96	74
Levels 2 and 3 ('information')	4	26

This pattern of findings was once again subject to statistical tests to assess the probability it arose owing to random error. In this case $\chi^2 = 18.98$; df = 1; p < 0.0001 which means that there is a significant association between the direction of communications (transmit or receive) and their type (data or information). The observed results differ significantly and meaningfully from the results expected under the null hypothesis (e.g., Cramers V = 0.31; p < 0.0001, a medium effect size).

Overall, what these findings seem to convey is that the incoming comms arrive on the digital layer as data and leave the BDE HQ on the voice layer as information. A possible design goal of future versions of the system might be to increase the proportion of information on the non-voice aspects of the NEC system. Of course, we were not anticipating 100 per cent of data to be turned into 100 per cent of information, but, that said, it is still interesting to note that the relatively simplistic voice layer achieved more of this than the highly complex digital layer (reinforcing the earlier point about simple interactions and complex tasks). The more pressing practical issue is that with bandwidth at such a premium in these systems, human-factors analysis could potentially provide a useful and legitimate criterion for judging how this resource should be most efficiently loaded. This issue is reminiscent of the kind of perceptual coding seen in some data compression techniques within the world of signal processing. In this case great economies in data-rates are achieved by only encoding those which are 'perceptually salient'. Likewise, could similar economies be achieved by encoding only those data that are 'cognitively salient'? To express the current situation in cruder terms, this is a lot of data (and a lot of bandwidth) for not much resultant 'information'.

In wider social network terms it is interesting to note that the behaviour of the live-NEC system as a whole seems to behave quite literally like an open system. Specifically, the pattern of receive and transmit events resembles 'a model of a biological organism' in that it 'receives [data] from its environment (intelligence), makes decisions, and produces some effect on its environment' (Dekker, 2002, p. 95). If NEC is analogous to an open system, as Figure 6.4 would seem to suggest, then it should be able to achieve steady states based on a continuous throughput of information (not merely data) and able to do so despite being subject to a range of external disturbances. It should also be able to adapt, with emergent properties like Free Text and the sort of information intensive interaction provided by the voice comms function enabling the system to build further structure and evolve itself to progressively better adapted states.

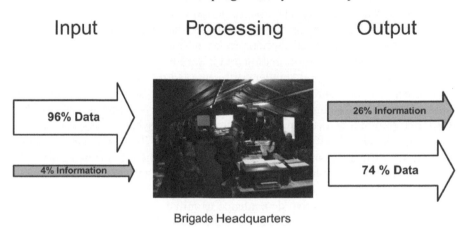

Input   Processing   Output

96% Data

4% Information

26% Information

74 % Data

Brigade Headquarters

Figure 6.4   The biological conception of NEC sees a constant throughput of information, processing, output, steady states and the ability to evolve and adapt

Sensitive Dependence on Initial Conditions

Emergent behaviours do not just rely on the boundless possibilities implied by complexity and open systems behaviour; they also rely on constraints. Emergent behaviour depends on at least a certain degree of coordination between the activities of the participating agents. The agents may themselves be complex but if they operate in a totally unconstrained way then 'no higher scale is available for the encompassing system'(Atay and Jost, 2004, p. 21). The amount of un-constraint is what pushes emergence from weak through strong, then onwards to complete randomness (as per Table 6.1). So if emergence cannot 'emerge' without some form of constraint, then how do complex systems behave under different constraints?

Another prominent artefact of complex systems is 'sensitive dependence on initial conditions'. The term often given to this phenomenon is 'the butterfly effect', in recognition of a paper by mathematician and meteorologist Edward Lorenz. The title of Lorenz's paper was 'Does the flap of a butterfly's wings in Brazil set off a tornado in Texas?' (see Hilborn, 2004). Lorenz argued that, at least in theory, it could. The supposition was based on the now legendary story of how a primitive computerised simulation of the weather gave birth to an entirely new scientific discipline, that of Chaos Theory. Midway through a weather simulation, some interesting phenomenon had emerged and the simulation was paused in order to look more closely. To avoid restarting the simulation from scratch, the state of the simulation at that moment, which was represented by a single number expressed to six decimal places (.506127), was re-entered but to only three decimal places (.506) in order to save time. The assumption was that any effects of such a small difference would be damped out by the larger and presumably more powerful global phenomenon. This assumption proved to be incorrect. .506127 represented the initial conditions of the simulation and it was highly sensitive to change. This one part in a thousand (and smaller) difference led the simulation, over a fairly short time, to evolve into a completely different state. So why the butterfly? It is because a flap of a butterfly's wings may represent a one part in a thousandth, millionth, billionth or even smaller of a total weather system, but given enough time it is conceivable (if not especially probable) that its presence and absence could be the difference concerning a tornado in Texas. The location of the butterfly and the ensuing tornado have varied over the years, but the fundamental attribute of 'sensitive dependence on initial conditions' has remained.

Like emergence, sensitive dependence on initial conditions is also a feature of human-factors problems and can be illustrated with recourse to the lab study presented in Chapter 3. Here, a simplified classic C2 organisation was set to work within a simulated environment and pitted against an NEC organisation on an identical task. Both organisations contained live actors and they performed a C2 task within a complex, adaptive environment. The traditional approach to examining the relative performance of these two command and control organisations might have been to garner a sample of teams, subject them to one or even a few practice trials, then expose them once to the experimental condition. Inferential statistics would then be applied and these findings would be inferred to the population from which the sample was drawn.

The presence of the butterfly effect and the organisational properties of NEC, however, create different experimental requirements and somewhat different experimental questions, questions that require such a study to go beyond traditional human-centred techniques and not rely on static representations of the human-system interaction (Lee, 2001; Woods and Dekker, 2000). The focus shifts to how the incumbents of the different organisations adapted themselves and their context to suit their needs and preferences, and how the initial conditions, or constraints, represented by the two organisational types (NEC and C2) affected this process over time. As a result, the situation shifts from running an 'experiment' to something more akin to running a 'model', albeit one with human participants. The benefits of this so-called micro-world approach are that it allows human adaptability to flourish within a dynamic 'interior environment' of a system. Another major advantage for military research is that this approach also allows novice participants to become highly expert in a task (which is an often criticised component of traditional cross-sectional studies which invariably rely on non-expert samples). In the present case, multiple regression was pressed into service as a means of time series analysis. The goal becomes one of being able to detect, statistically, the underlying theory behind the data. If the constraints of C2 and NEC represent initial conditions the question is what system results from sensitive dependence on them? Chapter 3 provided an answer.

The point behind sensitive dependence, one which is often ignored in human factors research, is the time dimension. For problems that reside at the 'complex' end of the problem space, instantiating the system in the form of an experiment may not be enough. The system has to be 'run': 'Running a system is the quickest, shortest, and only sure method to discern emergent structures latent in it. There are no shortcuts to actually 'expressing' a convoluted, non-linear equation to discover what it does'. In the case of sociotechnical systems, 'Too much of its behaviour is packed away' (Kelly, 1994, p. 13).

From Typology to Taxonomy

What Chapters 3 and 5 make evident is that one of the conceptual problems with NEC research, from a human-factors perspective, is that it is very difficult to apply the conventional experimental approach to it. The more one isolates key variables and controls for other extraneous factors, the less the phenomena of interest looks and behaves like 'real' NEC. Experimental control, therefore, frequently finds itself in conflict with ecological validity and the inevitable question that 'you may have established a relationship between x and y, but what about z?' Although many NEC questions are perfectly amenable to the traditional experimental approach, others are not.

In the course of matching the research 'approach' to the extant research 'problem', we have already alighted upon the NATO SAS-050 approach space (in Chapters 2 and 5). In some respects this represents a conceptual alternative to the brute force exploration of all possible combinations of factors potentially relevant to NEC. It is a form of top-down, systems level view of C2, a relatively simple (or simplified) method that tries to model complex organisations. This section looks again to complexity to establish the conceptual foundations for this type of approach where it is evident that, despite a paucity of overt cross-referencing, the NATO SAS-050 approach space has a long and useful legacy. The insights so derived help to set the model into a wider context and justify the extensions that have been performed on it.

As described already, the NATO SAS-050 approach space is based on viewing C2 as a fundamentally information processing endeavour, as a cyclical pattern of information collection, sense making and actions (NATO, 2007). Attached to this model are the multifarious variables that seem relevant to the functioning of it, variables drawn from fields as diverse as general systems theory, human factors, cognitive psychology and operational research (amongst many others) and which number more than 300 in total. Tacit in questions of the form '…but what about z?' is the presence of correlation among factors, and this expedient is represented by establishing those 3,000 conceptual linkages between the identified variables. As stated earlier, this interconnected web of C2 variables is referred to as a 'Reference Model'. At this point one could begin to explore the model with experimental techniques. At the local level such an approach has much in its favour in terms of establishing causal relationships between specific sub-sets of variables. However, at the global level it quickly becomes intractable.

An alternative method seems to be encapsulated by the NATO SAS-050 approach space. Here, the Reference Model is subject to analysis with systems engineering tools in order to reduce the number of variables in the model down to the critical dimensions along which command and control organisations fundamentally differ from each other. This process of 'dimension reduction' arrives at the now familiar 'three key factors that define the essence of [command and control]' (Alberts and Hayes, 2006, p. 74): allocation of decision rights, patterns of interaction and distribution of information.

In the former case of x versus y causality and the traditional experimental approach, questions of the form '…but what about z?' can only be answered by including 'z' in subsequent analyses, thus expanding exponentially the experimental conditions required to test for z's effects on x and y. In the latter case, the same question is answered by referring to the high-level model and the fact that 'z' (not to mention the rest of the alphabet) is already embedded within it. The two strategies represent quite distinct approaches. The former is bottom-up and reconstructs the behaviour of the whole NEC system from the individual cause and effect relationships thought to exist between its parts. The latter is top-down and reflects an interesting cybernetic principle: 'if all the variables are tightly coupled, and if you can truly manipulate one of them in all its freedoms, then you can indirectly control all of them' (Kelly, 1994, p. 121). The effect of this in the development of NATO SAS-050-like models is analogous and runs as follows: if the NATO SAS-050 reference model (the interlinked web of 300 C2 variables) is coupled (through those 3000+ links), and the three axes of the model represent the degrees of freedom that these principle output variables afford, then in measuring them you simultaneously, if indirectly, measure all the rest. It is unlikely to be anywhere as near as precise as this but something of that cybernetic flavour lies behind the generic process of dimension reduction. What it enables the analyst to do, given that predictive efficiency at the component level is not guaranteed (i.e., Free Text), or even especially tractable (i.e., as variables continue to be added), is to switch the analysis from bottom-up to top-down. Thus the behaviour of individual components are not measured in isolation, rather the product of those behaviours are measured as a whole. In sociotechnical systems like NEC these emergent behaviours are just as likely to *be* the system as its individual constituent behaviours.

Phase Spaces

In the 'sensitive dependence on initial conditions' example above, just one attribute of a complex system's behaviour at a time (e.g., task completion time) is plotted as a simple time series. Complex systems, according to the attribute of multiplicity, are specified by more than one variable. At the other extreme from a simple time series is a fully multidimensional space (i.e., a matrix) which effectively contains an axis for every parameter of a system, assuming that every parameter can be known. The space created by the intersection of these axes is formally known as a phase space. Within it, every possible state that a complex system can adopt can be plotted.

Phase spaces are derived from the work of physicist Willard Gibbs and date back to the turn of the century. They, too, are multi-dimensional spaces with the most well-known examples, such as the Lorenz attractor (Figure 6.5), being three dimensional just like the NATO SAS-050 model. Into this space every possible

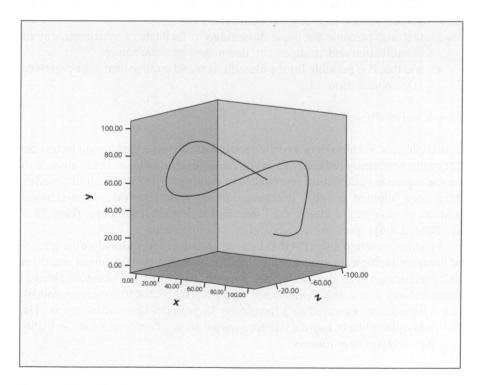

Figure 6.5   The Lorenz attractor: phase space showing the complex dynamical behaviour of a physical system

state that a physical or mechanical system adopts can be plotted as a unique set of coordinates. When these coordinates are linked by a solid line, the system in question traces a topographical pattern.

Phase spaces arise from the study of complex dynamical systems and the inadequacies of simple one-dimensional time-series analysis in capturing and representing their behaviour. Unlike simple time-series graphs, phase spaces are a graphical way in which to analyse the *interactions* between variables. Although the resulting representation is entirely abstract it can reveal fundamental properties about the system in question. Although abstract it is quite common for prominent graphical features to have some kind of physical analogue (in the Lorenz example it is fluid flow changing direction). In the case of the Lorenz attractor, three equations completely explained the behaviour of the physical system that traced its distinctive path through the phase space. The outputs of these equations enable the phase space axes to be expressed numerically. From the point of view of the NATO SAS-050 model and its antecedents, phase spaces reveal four critical insights:

1. that it is possible for complex dynamic systems (in this case physical systems) to be explained by a reduced set of simpler, more tractable equations;
2. that this top-down approach preserves the critical sets of interactions that give rise to the high-level system behaviour;
3. that it is possible for those dimensions to facilitate 'a systematic way of classification and arrangement' due to their numeric nature;
4. and that it is possible for the classification and arrangement to be presented in graphical form.

Functional Holography

Unlike physical systems, many of the variables of interest within human factors are not easily or usefully reduced to a set of fundamental equations. In the absence of formal equations alternative approaches to creating NATO SAS-050-like models have been adopted in other domains. Moving from physical and mechanical systems to biological systems, the Functional Holographic technique (Baruchi et al., 2004; 2006) represents the next level of sophistication.

Functional Holography (FH) has been developed as a way to analyse the activity of complex biological networks (a scientific euphemism for the human brain) via ECG recordings. The diffuse, highly dynamic pattern of activation obtained in these settings is not only reminiscent of the NATO SAS-050 Reference Model, but is normally represented as a formidably large multi-dimensional matrix. The FH technique subjects this dataset to a unique form of analysis which in highly summarised form is as follows:

1. The matrix of data is subject to initial evaluation and normalisation (see Baruchi, Towle and Ben-Jacob, 2005).

2. The dataset is then subject to Principal Components Analysis (PCA), the three leading eigenvectors (or clusters) forming three intersecting axes. By these means the N-dimensional space represented by the raw matrix is collapsed into a three-dimensional space, the three principal eigenvectors being assumed to explain the majority of the system's behaviour.

3. A comparison between pre and post normalised data (amongst other steps) enables live data to then be projected into the 3D space and visualised.

4. The plot of discrete points in the space are connected, interpolated and otherwise manipulated to form manifolds, curves and hulls in order to enhance the spatial/topographical features of the data.

5. New data can then be overlain across previous datasets to establish baseline conditions, the temporal ordering of activity, or a range of other comparisons.

Figure 6.6 shows the outputs of the FH technique where the similarity between it and the NATO SAS-050 approach space is readily apparent. From the point of

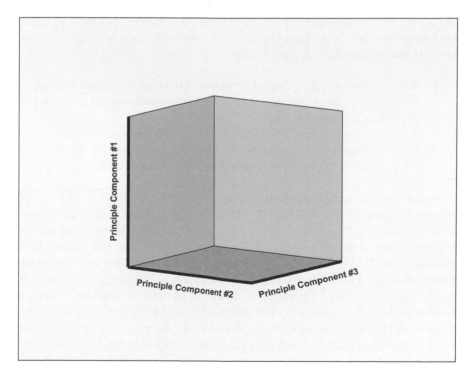

Figure 6.6   **Multidimensional space collapsed into three leading axes via PCA**

Source: Baruchi, Ben-Jacob and Towle, 2005b.

view of the NATO SAS-050 model, and its antecedents, FH reveals the following insights:

1. For complex dynamic systems that are not amenable to formal expression via mathematical equations, clustering techniques provide an alternative route to defining a reduced set of principal explanatory factors.
2. Manifolds, curves and hulls represent an advanced form of visualisation which could show how command and control organisations vary over function and time, just as the same methods enable different sets of ECG data to be compared dynamically.

To summarise, phase spaces transform a complex dynamical system into a set of equations to create a drastically reduced subset of key variables which explain the greater part of its behaviour. If the subset of variables is equal to three then the dynamic behaviour of the system can be plotted into a 3D space and be visualised. Functional holography lifts this approach out of the mathematical formalism inherent in physics and applies it to biological data. Instead of a system being specified by a set of formal equations, a reduced set of factors is revealed by PCA. The legacy of this convergent approach to modelling complex systems is a generic three stage method which the NATO SAS-050 approach space, and its subsequent development in Chapter 5, fits neatly into:

1. Stage 1: Reduce the complex system's behaviour to a small number of primary axes which describe a large (or at least meaningful) proportion of the system's total behaviour.
2. Stage 2: Turn the typological scales into numeric taxonomic scales.
3. Step 3: Measure live examples of the complex system in question and project them into the space created by the taxonomic scales.

Phase spaces and Functional Holography seem to provide a powerful retrospective legacy for the NATO SAS-050 approach space. The similarities between all three of these approaches are readily apparent. The maximal amount of information about an organisation's dynamical behaviour has already been reduced by the NATO SAS-050 approach space to three variables in a manner not dissimilar to the FH technique (at least conceptually). Mapping social network metrics onto these three variables (as described in detail in Chapter 5) turns it from a typology to a taxonomy. The enhanced NATO approach space, therefore, maps to, and is derived explicitly from, established concepts and methods found in complex systems research. From a human factors perspective the field of complex systems research provides a rich source of insight, of which this chapter barely scratches the surface.

Dynamical System Behaviour

The conceptual footholds established above provide an opportunity to look afresh upon the 'dynamics' of NEC. The periodogram approach used in Chapter 5 to look at tempo suggests that not only do such dynamics exist but that they behave in certain ways. What is visually evident is that the live-NEC system clusters into reasonably distinct regions of the approach space as if those regions possessed a kind of gravitational pull. If we are equating the NATO SAS-050 approach space with the notion of phase spaces, then such regions can be loosely termed 'attractors'. In the case of clusters that form in defined areas of the phase space, these can be termed 'fixed attractors'. Fixed attractors represent something analogous to equilibrium state(s) of the system into which it is repeatedly drawn, the prevailing behaviours when 'everything settles down'. This does not fully describe the dynamical behaviour of the live-NEC organisation, however, because many of the other points do not fall into defined clusters at all and are instead scattered widely. The organisation is thus attracted to these other points in the space by forces whose underlying dynamics are not stable and deterministic, but unstable and chaotic. As such, they could be labelled as 'strange attractors'.

The presence of attractors suggests a much more fundamental source of organisational dynamics which is related to the problem space itself. The fact that the organisation is drawn or otherwise propelled into different regions of the phase space indicates that, like a ball rolling across a surface containing dips and hollows, the environment possesses a defined 'causal texture' (e.g., Emery and Trist, 1978). This is a classic sociotechnical systems concept that maps back across to the study of complexity. The causal texture of environmental change provides a conceptual road map of attractors (strange and/or fixed) that reside in the approach space. Fixed attractors might reside in what are referred to as the 'placid random' and 'placid clustered environments'. In both these cases, the underlying environmental dynamics are fundamentally stable and therefore lend themselves to more rationally designed organisations such as classic C2. Strange attractors seem to reside in what are termed the 'disturbed reactive' and 'turbulent field' environments. In both cases, the underlying dynamics are more disorganised, complex, even chaotic, thus prompting the requirement for an organisation that can better track those sorts of dynamics, e.g., NEC. Of course, an organisation is itself part of the environment and through its actions able to influence its causal texture.

Summary

In this chapter we have attempted to further blend human factors, sociotechnical systems theory and complexity together. This we do in order to explain the 'why' of Chapter 5's live-NEC case study and to set the development of the NATO SAS-050 approach space into some kind of context.

We have discussed emergence, open systems behaviour, sensitive dependence on initial conditions, explained how the NATO SAS-050 approach space has been extended from typology to taxonomy, and considered the causal texture of environmental change. As mentioned above, this is in no way an exhaustive list and we barely scratch the surface of the insights that complex systems research can provide for this type of human factors problem. But as a pointer towards the shape of future insights we see the following.

The synergistic (and anti-synergistic) effects of emergence clearly go beyond the simple expedient of 'things being greater/or less than the sum of their parts'. There is a clear and coherent typology, and via Relative Predictive Efficiency (RPE), a conceptual inroad into actually testing for it. The value in diagnosing emergence is that it can help to define the proportion of the problem space that can be described as merely 'complicated' (and for which traditional deterministic ergonomic methodologies are most useful) compared to those that are 'complex' (for which alternative ergonomic approaches start to become justified).

The type of experimental paradigm that dominates human factors is the one factor psychology experiment, in which variables are isolated, the context controlled, and the human participant's behaviour assessed (typically once) in the control versus treatment conditions. If we assume all human factors problems to be located in the part of the problem space bounded by stability, certainty and complete knowledge then this represents the ideal approach. If, however, putative human factors problems (or parts thereof) can be characterised by instability, uncertainty and lack of knowledge, then the methodological strategy has to shift from 'experimental control' to the set of 'initial conditions' from which humans are permitted to evolve from. As such, we go from cross-sectional studies to longitudinal ones, from experiments to micro-worlds. Neither approach on its own is adequate in all circumstances; it is a question of matching approaches to problems.

A long-standing strategy for modelling complex systems is to preserve the underlying interactions and reduce the number of variables that explain a system down to the few that explain the majority of its behaviour. Exactly this approach has been undertaken for physical systems since the turn of the century and the extended NATO SAS-050 model of command and control represents the latest in a long line of similar 'dimensionally reduced' models.

Chapter 7
Beyond NEC

Aims of the Chapter

In this chapter we remain at the level of the total system. Here we explore exactly what sort of system a sociotechnical system is. Is it as synonymous with NEC as the literature and studies might suggest? How does NEC compare to other real-life organisations that are designed to meet similar challenges? This chapter sets out to answer systems-level questions of this sort.

Having extended the NATO SAS-050 approach space, couched it within a wider context of complex-systems research and subjected it to theoretical and practical application, an opportunity arises to now use it as a way of understanding explicitly what it means, organisationally, to occupy different regions of the approach space. To this end, a diverse collection of organisations, from theoretical archetypes to terrorist networks, were modelled with social network analysis and their position within the enhanced NATO SAS-050 approach space fixed. The supporting evidence and research data appended to these organisations was spun in and used to further our understanding of what sort of organisation live-NEC actually is, could, and perhaps 'should' be.

The results of this benchmarking exercise reveal a fairly distinct region of the NATO SAS-050 approach space into which many real-life organisations, live-NEC included, seem to gravitate. This region is, in turn, associated with the attributes of 'small-world networks'. To the extent that this network topology is relevant to NEC (and there is good evidence that it might be) then we see sociotechnical systems theory emerging as an explicit strategy for creating such topologies.

The Ultimate NEC

The techno-organisational vision of NEC has been described as, '...self-synchronizing forces that can work together to adapt to a changing environment, and to develop a shared view of how best to employ force and effect to defeat the enemy. This vision removes traditional command hierarchies and empowers individual units to interpret the broad command intent and evolve a flexible execution strategy with their peers' (Ferbrache, 2005, p. 104). Taken to its limit, the distribution of information required to develop 'shared views', the allocation of decision rights congruent with 'empowering individual units' and the patterns of interaction required for 'flexible execution strategies' calls forth an image, if not of Wal Mart (e.g., Shachtman, 2007), then of a 'fully connected information-age system' (Alberts and Hayes, 2005, p. 91). Such a system, the so-called Edge Organisation, is a topological maximum represented by every agent/node/member of military personnel connected to every one else (Alberts and Hayes, 2005). Of course, '...the actual number of interactions that need to take place is not the theoretically possible maximum' (2005, p. 91), in which case, what is it? Under the rubric of 'shared views', 'flexible execution strategies' and the 'empowerment of individual units' what organisational topology are we speaking of?

Various authors hint at the possibility that, in actual fact, NEC could/should/ will behave as per something like the internet, with communities of interest, hubs of high connectivity and networks in which the strength comes from 'weak ties' (e.g., Granovetter, 1973). An opportunity arises in this chapter to contribute meaningfully to this debate from a uniquely human centred perspective.

Dimensions of Organisation Structure

The Aston Studies

To begin this study of cutting edge organisational design, of which NEC is an example, it is useful to go back 40 years to the eponymous Aston Studies. The aim of these studies, the effect of which is still felt in the field of organisational science, overlaps with ours. They, too, were venturing towards 'more sophisticated conceptual and methodological tools [...] for dealing systematically with variations between organisations' (Pugh et al., 1968).

The Aston Studies began by surveying 52 different work organisations, from multi-thousand employee public corporations to small family firms, on 64 dimensions which the then extant literature identified as being of importance (in terms of how organisations differ from each other). These 64 dimensions were subject to evaluation, normalisation and principal components analysis (as in the Functional Holographic method described above in Chapter 6) in order to reveal a reduced subset of underlying factors (as in Phase Spaces, also described in Chapter 6). The Aston Studies revealed three basic dimensions of organisational structure defined by Pugh, Hickson and Hinings (1969) as follows:

Factor I – Structuring of Activities

'...the degree to which the behaviour of employees was overtly defined, incorporating the degree of role specialization in task allocation, the degree of standardization of organisational routines, and the degree of formalization of written procedures.' (p. 116). In the language of sociotechnical systems theory, this factor clearly relates to the degree of 'critical specification'. At one end of the spectrum is 'minimal critical specification' or 'Effects Based Operations', which focuses on outcomes, or 'maximal critical specification' or 'Action Based Operations' in which the means to achieve an outcome are prescribed in detail.

Factor II – Concentration of Authority

'...the degree to which authority for decisions rested in controlling units outside the organisation and was centralized at the higher hierarchical levels within it' (p. 116). In the language of sociotechnical systems theory this factor describes the extent to which roles within the organisation are rationalized and at what level.

Factor III – Line Control of Workflow

'...the degree to which control was exercised by line personnel as against its exercise through impersonal procedures' (p. 116). Again, in sociotechnical terms this describes the autonomy granted to personnel, with 'semi autonomous work groups' at one end of the spectrum to relatively undifferentiated, prescriptive job roles at the other.

There is little evidence that the NATO SAS-050 approach space has referenced this legacy work (it is not mentioned in the reference list to NATO, 2007, Alberts and Hayes, 2005 or Smith, 2006) which in one sense appears surprising given that Figure 7.1 emerges as being virtually identical to the NATO SAS-050 approach space. On the other hand, this apparent lack of cross-referencing can be seen as an advantage, because despite the forty year time interval, the NATO SAS-050

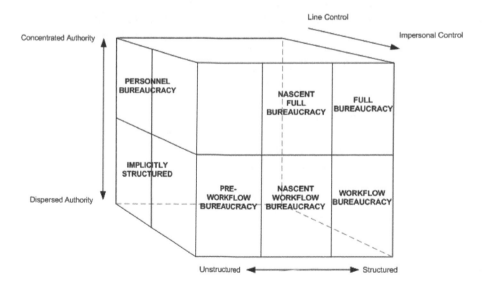

Figure 7.1   The spectrum of organisation types generated by the Aston Studies are a 3D cube conceptually very similar to the NATO SAS-050 approach space

Source: Pugh et al., 1969, p. 123.

approach space seems to have converged, independently and by different means, on a very similar set of primary organisational factors:

- Structuring of activities relates to distribution of information (the degree to which behaviour is overtly defined relates strongly to the distribution of information within an organisation).
- Concentration of authority has a clear link to allocation of decision rights.
- Line control of work flow relates strongly to patterns of interaction.

Unlike the NATO SAS-050 approach space, the Aston Study's 'approach space':

- *is* a taxonomic approach, using quantitatively defined axes in order to systematically plot an organisation into the space;
- *is* an attempt to plot where an organisation actually resides in this space, as opposed to focusing on where it thinks it is, or where it should be;
- *is* based on a systematic (and substantial) analysis of live organisations rather than theoretically defined ones.

The Aston Studies complement the NATO SAS-050 approach space by:

- showing that a typological model of organisations *can* be rendered as a taxonomic model (providing suitable measures and metrics can be defined);
- showing in turn that the Phase Space and Functional Holography techniques *can* be made to apply to entire organisations, in concept if not in detail;
- showing that the NATO SAS-050 model has converged on a *common set* of three primary factors arrived at independently, by different means and in different organisational settings.

The Aston Studies do for organisations what Functional Holography does for biological data. They also use PCA to derive a three-factor model of organisations, enabling any organisation to be plotted into a 3D space. If 40 years ago the Aston Studies populated their 3D model with live data from a diverse set of organisations then the same can be done with the enhanced NATO SAS-050 approach space. This should help to situate the live-NEC case study described in the previous two chapters within some sort of organisational context and show precisely what sort of organisation it actually is.

Beyond NEC

Crossing the Divide

We are now ready to populate the NATO SAS-050 approach space with a wide variety of real-life organisations, including NEC. It is fair to say that comparing military and civilian organisations all too frequently yields the objection that civilian organisations are simply 'too different' to NEC which therefore renders any such comparisons meaningless. This objection needs to be briefly tackled.

There are at least four objections to excessive militaristic insularism. Firstly, although there are indeed key differences either side of the military/civilian divide, there are also key similarities and lessons to learn from the often considerable legacy of research that is appended to non-military domains (the Aston Studies are a case in point). Secondly, many civilian organisations are already in areas of the approach space that NEC might aspire to. It would be remiss to not to look at them and seek out opportunities to spin-in insights and lessons learnt. Thirdly, in an era of increased multi-agency working and interoperability it seems entirely pragmatic to look at differences/similarities between various organisations in order to facilitate effective command arrangements between them. Fourthly, and by no means finally, the opponents to which NEC is a conceptual response are often distinctly non-militaristic in nature (at least in the traditional sense). The concept of 'organisation' and of 'command and control' is in some cases supplanted by 'informal' organisations with a distinct 'lack of control'. This on its own creates a need to look open mindedly at the essential nature of 'approaches' and 'problems'.

The practical response to objections of this sort is an analytical progression from network archetypes, commercial organisations and civilian examples of command and control to conclude with the central topic of this chapter and a look at what lies 'beyond NEC'. This is achieved by analysing innovative theoretical and actual organisations (from terror organisations to the internet) that seem in their own distinct ways to not only be coping with complexity but motivating the development of NEC itself. This in turn gives cause to revisit the stereotype of an edge organisation and to consider instead some important and long standing variations based (implicitly or otherwise) on sociotechnical principles. In a sense, then, the analysis will be going full circle, from social network archetypes contemporaneous with sociotechnical systems theory, forward in time through the Aston Studies and formal civilian organisations, through next generation (internet-type) organisations, back to sociotechnical organisations. The analysis begins with NATO SAS-050 derived organisational archetypes, the 'Edge Organisation' and 'classic C2'.

NATO SAS-050 Derived Archetypes

Archetypes

In various contemporary readings on NEC a distinction is made between concepts such as 'classic C2' (i.e., the traditional, hierarchically arranged form of 'industrial age' command and control) and so-called 'edge organisations' (i.e., the distributed, highly interconnected form of information-age command and control given the label NEC). For the purposes of the current analysis we put forward the following 'Edge Organisation' and 'classic C2' archetypes as shown in Figure 7.2.

Classic C2   Edge Organisation

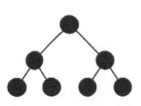

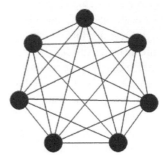

Figure 7.2   Archetypal networks representing Classic C2 and Edge Organisation

Alberts and Hayes, the major proponents of these two archetypes, are quick to point out that: 'The simple Industrial Age hierarchy appears to have very few connections when compared with a fully connected information-age system with the same number of nodes. Hence, naïve analysts conclude that the bandwidth requirements for a robustly networked force must, by definition, be massive and supportable only at enormous cost' (2005, p. 91). As they are quick to acknowledge, this is incorrect. The Edge Organisation represents more a theoretical maximum than a practical reality. What it serves to highlight is the potential for interaction, that networks can grow by evolution and acquire the ability to self-organise and adopt the structure required to match an extant problem. We will return to this issue throughout the analysis.

NEC Benchmarked Against Archetypes

Figure 7.3 shows how the normalised social network metrics discussed above have been used to populate the NATO SAS-050 approach space with Leavitt's (1951) archetypes from Chapter 3, along with Alberts and Hayes (2005) archetypes (shown above). It is possible to discern some similarity between the Edge and Circle archetypes, at least in terms of the networks being fixed within broadly the same 'octant' of the cube (i.e., octant #6). The lack of maximal interconnection in the case of the circle archetype clearly moves it further down both the distribution of information and patterns of interaction axes. Based on what is known about these archetypes, it can be hypothesised that organisations which find themselves plotted into this octant of the space can be associated with the 'active, leaderless, unorganised, erratic, yet enjoyed by their members' characteristics of highly networked 'networks'. This data driven finding supports the suppositions made of the NATO SAS-050 model and the classic C2 and Edge Organisations plotted into it (e.g., Smith, 2006; Alberts and Hayes, 2005; NATO, 2007).

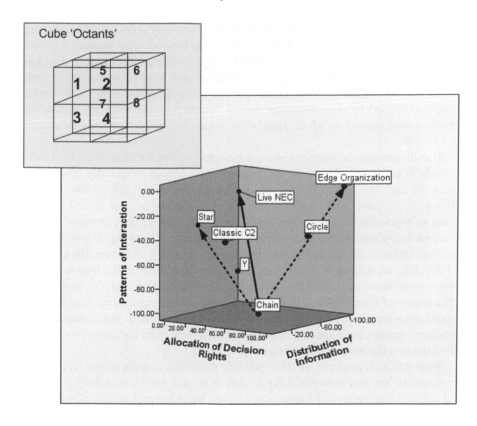

Figure 7.3   NATO SAS-050 approach space populated with network archetypes and a live instance of NEC. Illustration also shows NATO SAS-050 model divided up into eight 'octants' and the actual versus anticipated trajectory of a live-NEC organisation

In the diagonal octant (octant #4) lies the chain archetype, the Y archetype, and the classic C2 archetype. These networks can be characterised as being 'less active, possessing a reasonably distinct leader, well and stably organised, less erratic, yet less satisfying to most of its members'. Octant #4 contains organisations that are able to satisfy certain key performance indicators of deterministic problems, displaying a high degree of efficiency, yet are perhaps not jointly optimised in the sociotechnical sense given that these are organisations which seem not to be especially enjoyable to work in, nor perhaps to stimulate team cohesion.

There appears to be a trajectory running from octant #4 and the chain archetype, diagonally up to octant #1 and the Star network. This highly centralised network clearly hits a theoretical maximum in terms of their being unitary decision rights in the literal sense of there only being one person at the centre of the network. This clearly has an effect on distribution of information, which measures low as a result of 'the peripheral agents merely funnelling information to the centre'. This network, in effect, is a cluster, and networks exhibiting this property seem to fall somewhere in the mid-region of patterns of interaction. This network also measured lowest for 'enjoyment' in Leavitt's study (1951), suggesting that allocation of decision rights plays an important role in team cohesion and joint optimisation. In fact, unitary decision rights could be regarded as the antithesis of the 'semi-autonomous groups' espoused by sociotechnical theory. Octant #1, therefore, emerges as an organisational region in which jointly optimised organisational structures seem not to reside.

Also discernable is another tentative trajectory springing again from the Chain network in octant #4. This time the trajectory progresses diagonally upwards to octant #6 and the edge organisation archetype. Approximately midway along this line is the circle archetype (which seems to straddle octants #2 and 6). This seems to represent an alternative course of organisational development. Instead of increasingly centralised bureaucracies, the trajectory heads up in the opposite direction towards increased decentralisation. Interestingly, however, the example of NEC falls on neither of these trajectories. The aspirations of the field trial were to move command and control from where it was thought to reside (i.e., somewhere in the region of the classic C2 archetype) towards something approaching the edge organisation archetype. Interestingly, it behaved in an unanticipated manner and finds itself occupying a space that is not only different from the aspiration but also off of the two theoretical trajectories just described.

What this live example of NEC actually possesses is some of the 'clustering' properties of the star network. Like a 'star' it locates itself somewhere along the mid point of the patterns of interaction axis and leans towards unitary allocation of decision rights. However, it shares with the circle network broader dissemination of information. By populating the NATO SAS-050 approach space with an increasingly diverse range of live and theoretical organisations, some light on the unique properties of networks in this region can perhaps be shed.

NEC versus Commercial Organisations

Commercial Organisations

In the late 1960s the collective body of research now known as the 'Aston Studies' met a similar goal to that which NATO SAS-050 has recently been contending; to identify the critical subset of structural variables that explain the maximum amount of variation in the way that organisations differ from each other. The Aston Studies were based on a study of 52 organisations drawn from the 293 employing units then operating in the West Midlands area (and although not cited by name they clearly include many of the famous confectionery and car makers that the area is famous for). 'In ownership, the sample ranged from independent family-dominated firms to companies owned by private shareholders, a cooperative, branches of factories or large organisations, municipal departments, and national organisations'

(Pugh et al., 1968, p. 67). In size, the sample ranged from approximately 250 employees to in excess of 2,000.

Although largely historical in nature, this data serves a purpose in providing a substantial counterpoint for the types of commercial organisations that have come since. In a sense, they are to commercial organisations what classic C2 might be to NEC. However, NEC is not merely about computing and networked technology and these two aspects are not necessary precursors to being able to occupy distinctly 'edge organisation' regions of the approach space, as will be noted shortly. The legacy of the Aston Studies is able to make a much more proximal contribution to the NATO SAS-050 approach space in the form of an empirical taxonomy that provides a set of labels and attributes for at least some of the approach space's eight octants. Although phrased in the lexicon of 1960s organisational theory, it is not hard to see the crossover to military organisations. The taxonomic definitions are as follows.

Implicitly Structured Organisations

This organisation type is at the opposite extreme from the full bureaucracy:

> 'It might be thought that this minimal structuring [broad dissemination of information] and dispersed authority [decision rights] suggests unregulated chaos. [...] They cannot be labelled unstructured; for their structure, as far as the measures used go, is probably implicit. [...] these organisations are run not by explicit regulation but by implicitly transmitted custom, such as the traditional means usually typical of organisations of small or medium size where ownership and management still overlap. [...] These implicitly structured organisations are comparatively small factories (within the size range of the sample); they tend to be independent of external links; and they have scores on concentration of ownership that indicate that the operational control of the organisation has remained with the owning directors.' (Pugh et al., 1969, p. 118)

Personnel Bureaucracies

> '...organisations that show a high concentration of authority [unitary decision rights] and low structuring of activities [broad dissemination of information]. The authority of these organisations is centralized [...] and in many cases, such organisations do not structure daily work activities very much. [...] however, they have central recruiting, selecting, disciplining and dismissing procedures, conducted by formally constituted boards, and official establishment figures, appeal procedures, and the like. [Personnel bureaucracy's] bureaucratize everything related to employment, but not the daily work activity to the same degree. [They are] typically local or central government departments (for example, a municipal education department or the regional division of a government ministry) and the smaller branch factories of large corporations.' (Pugh et al., 1969, p. 117)

They are a hybrid mixture of bureaucracy and freedom of action. An interesting contingent response.

Workflow Bureaucracies

> '...the organisations that are highly structured [tightly controlled dissemination of information] and have a low concentration of authority [devolved decision rights] are found to be large, mainly independent, and to have their workflow relatively highly integrated, as in the vehicle components, standard engineering, foodstuffs, confectionary, and rubber goods factories, with their production schedules, quality inspection procedures, records of output per worker and per machine, records of maintenance and so on. This particular structural combination is that developed by large-scale manufacturing, or big business.' (Pugh et al., 1969, p. 117)

Full Bureaucracies

> 'This organisation [shows] the characteristics of a work flow bureaucracy (for example, standardization of task control procedures), as in large manufacturing corporations, together with the characteristics of personnel bureaucracies (for example, centralized decision making) as in government departments. The organisation is a manufacturing branch factory of a central government department.' (Pugh et al., 1969, p. 118)

The interesting point to make here is that this stereotype of a truly industrial age organisation (as in the literal Victorian sense of the term) seems a rare thing indeed, even 40 years ago. Clearly, commercial organisations, like command and control organisations, tend to fall well between such extremes.

Organisational Evolution

Briefly, it is worthwhile considering, as Pugh et al. (1969) do, that the octants of the approach space which these taxonomic classifications provide labels for imply some sort of developmental sequence. This sequence runs from implicitly structured organisations (of small size), which evolve into workflow bureaucracies (and the factory principles of the production line and adding value to raw materials) to personnel bureaucracies (in which the commodity being dealt with is more informational in nature). NEC, and other modern organisational paradigms, seems to make this developmental cycle go full circle by trying to recapture, within large organisations, some of the properties of small organisations, such as implicit structuring and the advantages that Leavitt in Chapter 3 notes in terms of activity levels and change.

NEC Benchmarked against Commercial Organisations

Figure 7.4 presents the approach space labelled according to Pugh et al.'s (1969) taxonomic definitions. It is clear that not all regions in the approach space can be specified by this taxonomy; however, the live instance of NEC does fit the classification of a 'personnel bureaucracy'. This indicates clearly that it is some way, conceptually and descriptively, from the classic C2 archetype against which it is most often compared. Rather than classic C2 and a Full/Workflow Bureaucracy, a Personnel Bureaucracy connotes a type of organisation made up of professionals in which control is exercised by 'line personnel' as against impersonal procedures. Translated into the language of military NEC, what we see here is evidence of

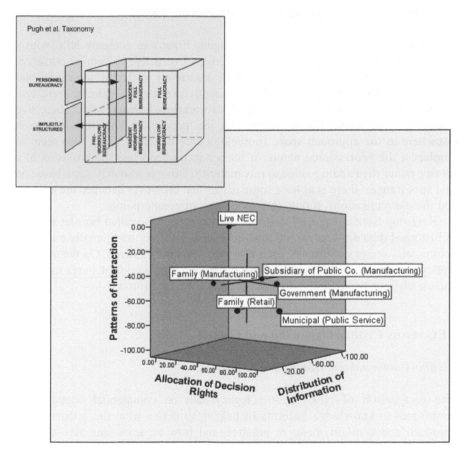

Figure 7.4   **NATO SAS-050 approach space populated with commercial organisations from the Aston Studies, a live instance of NEC, as well as being divided up into the empirical taxonomy offered by Pugh et al. 1969**

effects-based operations, or in the language of sociotechnical theory, minimal critical specification.

The data used to populate Figure 7.4 is based on averaging across commercial sectors which, as a result, tends to mask the individual organisations that, like NEC, can also be regarded as personnel bureaucracies. In more detail these are:

- government inspection department;
- cooperative chain of retail stores;
- local authority education department;
- savings bank;
- local authority civil engineering department;
- food manufacturer;
- local authority water department (Pugh et al., 1969).

Whilst it seems at face value to be slightly bizarre to compare NEC with a cooperative chain of retail stores, it is perhaps instructive to remind ourselves that these emerge as organisations with manifestly different structures and problems to deal with than, say, vehicle manufacturers, glass manufacturers, tyre manufacturers, or all the other examples of workflow bureaucracies. These live examples of personnel bureaucracies lack the 'production line' principles seen elsewhere in the approach space (notable octants #3 and 4); they also seem to emphasise the professional nature of the job incumbents and the provision of a service rather than adding value to raw materials. Strange as it may seem based on first appearances, there is at least some conceptual crossover between the military and these organisations, if only to provide a set of counterpoints.

Referring back to Figure 7.4, if we have to characterise rather broader swathes of historical data on what would now be regarded as very traditionally organised commercial sector companies, then it is in the manner depicted. On the average, legacy organisations such as these are predominantly 'workflow bureaucracies', and clearly, our live example of NEC is quite different from this.

NEC versus Civilian Command and Control

Civilian Command and Control Organisations

The comparison of NEC against legacy data on commercial organisations contributes to knowledge in terms of helping to define what the octants of the approach space might mean in practice, and how organisations may develop over time and with increasing size. These insights can be further enhanced by populating the approach space with several examples of much more relevant data from the civilian sector. The examples in this section are all versions of civilian command and control.

Air Traffic Control

Data for this organisational endeavour is drawn from a previous study (Walker et al., 2008d). The analysis was based on live observation of a number of air-traffic control scenarios, therefore, the resultant social network, like that for the live-NEC case study, is a representation of where the organisation actually is in the approach space rather than where it might formally place itself.

For context, Air Traffic Control in the UK is provided by National Air Traffic Services (NATS). The highest priority for the organisation is of course safety, but it is also important for the service to facilitate the expeditious flow of aircraft in order that they can travel economically and arrive promptly. NATS has a number of centres around the country where control of air traffic is conducted, including the London Terminal Control Centre at West Drayton, Middlesex, at which the current analysis was performed. UK controlled airspace is divided into sectors, each of which is monitored by an air-traffic control team. As an aircraft travels through these sectors, responsibility for controlling it transfers from one controller to another. Making sure that aircraft pass through this airspace and take off and land safely is the key responsibility of controllers and the collection of people and technology within which they are embedded.

A social network was created from live observation of air traffic controllers at their work stations, data being drawn principally from transcribing the set of interactions occurring via radio and in person. Figure 7.5 shows the network that was derived. Further details of the scenario and other analyses performed on the data can be sourced from Walker et al. (2008d).

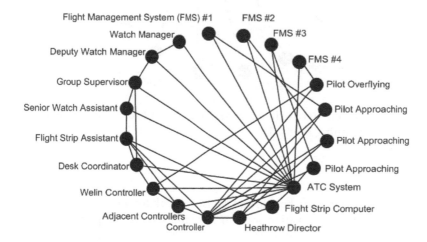

Figure 7.5   Social network derived from live observation of air traffic control scenario

Fire Service

Data to generate an appropriate social network, and to position that network within the approach space, was once again derived from prior work (Baber et al., 2004). Again, rather than produce a network that represents the formalised structure 'thought' to exist, the focus was on the structure of interactions that actually occur during the response to an incident.

The scenario modelled was based on a classroom training exercise undertaken at the Fire Service Training College. It involved qualified fire fighters and commanders talking through their response to a road-traffic accident scenario. From this data, the pattern of interactions in such an incident was gathered and it appears as in Figure 7.6. The data represents what the personnel (who have had experience of such incidents) would actually do, who they would speak to and what information they would require. Full information on the details of the scenario and other analyses performed on it can be sourced from Baber et al. (2004).

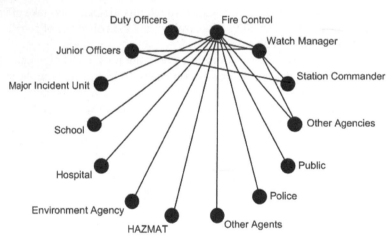

Figure 7.6   Social network derived from live observation of fire service training scenario

Police Service

The data used to generate the relevant social network for the police service is once again based on live observation and subject to much more comprehensive analysis within the relevant publication (Baber et al., 2004). The scenario is the chain of communications and interactions that follow a call from a member of the public in connection with a suspected vehicle break in. The social network so derived appears in Figure 7.7.

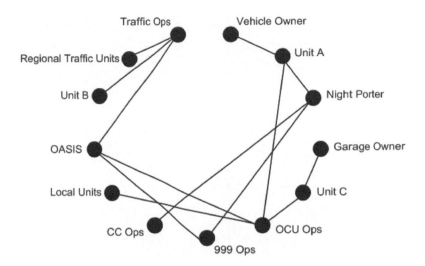

Figure 7.7 **Social network derived from live observation of police service incident response scenario**

NEC Benchmarked Against Civilian Command and Control

Air traffic control, police and fire services represent contemporary and well-developed civilian examples of command and control. In all cases, and perhaps unlike military operations in many respects, they are supported by a substantial information infrastructure. In the case of air traffic control every aircraft has a flight management system (FMS) which communicates with an air traffic control computer and it is this which combines with the ongoing dialogue between controllers and pilots to enable traffic to be controlled with a high degree of precision. Likewise, the fire service have a comprehensive radio and communications system, and the police service's suite of databases and other computerised facilities fall under a particularly advanced system called OASIS. To be fair, such systems work because a pervasive information infrastructure can physically be put in place (something that military forces frequently do not have the luxury of) and in the case of air-traffic control in particular, the problem to which the organisation is directed is undeniably complicated but it does at least obey fundamentally deterministic laws in normal operation (i.e., an aircraft is not likely to stop in mid-air and start flying backwards).

Police and fire services, on the other hand, have to deal with a great deal of unpredictability. The purpose of populating the approach space with these organisations is perhaps to reveal some of the advantages that NEC-like technology can yield under favourable circumstances. In cases in which the military and civilian agencies have to interoperate, it is of further interest to see if, and to what extent, these cooperating organisations fundamentally differ from each other. Figure 7.8

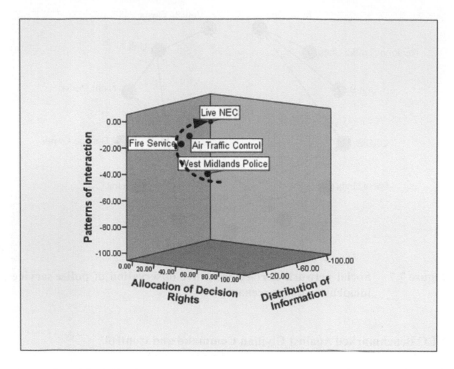

Figure 7.8   NATO SAS-050 approach space populated with civilian command and control organisations and a live instance of NEC

presents the results of this analysis where it can be seen that civilian command and control represents a shift away from the previous, rather bureaucratic organisations described in the section above, and in fact are reasonably well collocated with military NEC.

Whilst the comparison with commercial firms may be equally relevant to the benchmarking of NEC, the case for comparing these altogether more up to date and congruent civilian command and control organisations is much more easily made. What is interesting here is that we have found some contemporary and meaningful organisations that plot into roughly the same area of the approach space as the live-NEC case study (octant #5, or the 'personnel bureaucracy').

Once again, these examples do not plot into a region that is anywhere near the edge archetype (located in octant #6/implicitly structured). Indeed, there seems to be a quite distinct area of the cube into which these organisations gravitate (somewhere in the region of octants #1 and 5). What we can say that area is that it is characterised by a fairly high degree of unitary decision rights, yet with correspondingly broad levels of information dissemination and distributed patterns of interaction. The question to ask, therefore, is what is so special about this particular region of the approach space?

NEC versus 'Next Generation' Organisations

The Internet

Alberts and Hayes (2005) note in connection to the edge organisation concept that 'richly connected systems, such as the internet, are highly bandwidth-efficient because the actual number of interactions that take place is not the theoretical maximum, but is instead organized around communities of interest and is driven by circumstances' (p. 91). Despite attempts at internet cartography, no one really knows the full structure of the internet except to say that it does exhibit some quite distinct characteristics which enable reasonable estimates to be made (Ball, 2004; Dodge and Kitchin, 2001). The current internet topology is not the only network topology available, as Figure 7.9 shows.

The first option might be the centralised network topology. Early network paradigms were based on centralised processing and so-called 'dumb-terminals' (the 'centralised' topology shown on the far left of Figure 7.9). As such a network grows, and mirroring the same phenomena in social networks of the 'Star' variety, the central node quickly becomes overwhelmed. The network is also vulnerable. All it takes to completely disable it is to remove the central node.

Fully Connected Networks: at the other end of the spectrum from the centralised topology is the network equivalent of the edge organisation, in which

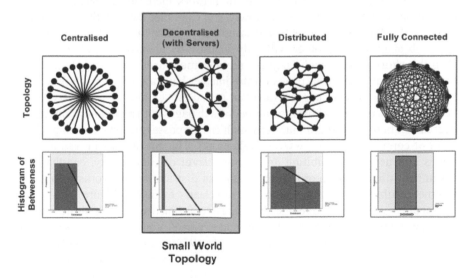

Figure 7.9 **Alternative network topologies with their corresponding histograms showing the characteristic spread of nodes having high and low levels of connectivity. The 'decentralised' network (as shown) is an example of a 'small-world network'**

every node (or in this case computer) is connected to every other one. This is similarly impractical in bandwidth and resource terms for reasons already alluded to (the cost of providing this level of redundancy is prohibitive). In between lies a region containing several other alternatives.

Distributed Networks: in appearance, this architecture looks a little like a city street plan organised along grid principles, with each junction representing a cross-roads. A distinct property of this network architecture is that as the network grows, so does the average path length between one node and another. In this case words, the number of steps to get from one point in the network increases more or less linearly with the network's size. Given that the internet, for example, is forecast to grow in size by 1000 per cent in the next few years, this sort of growth would be accompanied by a more or less 1000 per cent increase in the distance (and number of intermediate servers/computers) that would be required to reach a desired web page, accompanied by a corresponding increase in time to negotiate such a complex route (Ball, 2004).

Decentralised network (with local servers): another alternative, the one which seems to have evolved in practice, is the 'decentralised' network with 'local servers'. This topology is rather like the hybrid structures we have alluded to several times already (e.g., Chapter 3) in that they combine the high clustering of a centralised architecture without the vulnerability of having a single central computer. This feature arises because the hubs of these decentralised networks are in themselves quite richly interconnected. The network has an abundance of short cuts rather like bypasses and expressways that provide shortcuts across grid-like street plan. This feature means that as the network increases in size, the average path distance increases only very slowly. For example, faced with the internet's forecasted 1000 per cent growth, and assuming this structure, the distance between one web page and any other will only increase by about two steps. Combined with this short average path length is the aforementioned clustering. In practice, this means that many web pages on the internet have a comparatively low level of connectivity to other pages. However, as you move along an imaginary axis it becomes apparent that whilst the number of web pages with more interconnectivity decreases, there are a small number which end up possessing an exceedingly high level of connectivity (e.g., www.google.co.uk). This means that instead of most nodes in a network exhibiting an 'average' level of connectivity, one that tails off in both directions according to a Gaussian distribution, networks such as the internet exhibit a power function of the sort clearly evident in Figure 7.9.

Small-worlds: short, average path length combined with high clustering and the power-law relationship between nodes versus connectivity; these are all attributes of so-called 'small-world' networks (Watts and Strogatz, 1998). Small-world networks seem to represent a transition region between networks that are fully ordered (with a uniform level of connectivity for every node, as in Edge Organisations and fully connected networks) and fully random (those with a non-uniform level of connectivity derived from a random linkage/network re-wiring algorithm). In between this network-based order and chaos, in a clear analogue with many other complex phenomena, is a comparatively broad region in which the properties of networks change in quite specific ways. Firstly, moving from ordered to random networks, what happens is that average path length decreases sharply initially, then levels off. Clustering, meanwhile, does the reverse, remaining in a highly clustered state in the face of random re-wiring until it too suddenly drops in level, but much later on. Small-world networks come in a number of different sub-forms that span this region (e.g., Amaral et al., 2000). An archetypal small-world network is shown below in Figure 7.10 for illustration.

The small-world topology has attracted a great deal of attention in recent years because it has been found to occur in many natural as well as artificial networks. This includes the worm Caenorhabditis elegans (the only 'brain' that has so far been mapped out in its entirety) as well as the internet. Organisationally, if the bureaucratic classic C2 archetype fits the metaphor of a 'machine' (e.g., Morgan, 1986), and the aspirations of NEC are more associated with the metaphor of an 'organism', then something relevant and potentially useful may arise from considering small-world networks.

A long-standing finding in organisational theory is that organisations with these rather more organic characteristics are the ones that are able to adapt and perform well in turbulent environments. Kanter (1983) provides many real-world examples of where this has demonstratively been the case. As for small-world networks, they too have been associated with 'enhanced signal propagation speed, computational power, and synchronizability' (Watts and Strogatz, 1998, p. 440) all of which seem highly desirable characteristics for NEC. It is, however, important to point out that like all such topologies it is not ideal in all circumstances. Indeed, whilst there has been much enthusiasm for small-world networks since Watts and

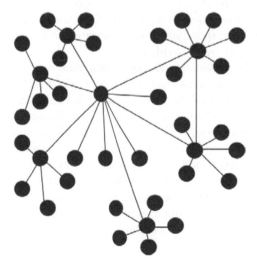

Figure 7.10 **The small-world network archetype revealing the topological features associated with high clustering and a short average path length**

Strogatz's innovations in this area there are disadvantage. For example, small-world networks do not guarantee synchronizability (e.g., Barahona and Pecora, 2002). In organisations they rely on people with unusually high connectivity (e.g., Gladwell, 2000), they still contain vulnerabilities compared to other network topologies if hubs can be identified and disabled (e.g., Ball, 2004), and one person's community of interest is another person's clique, with a set of associated maladaptive team attributes to accompany it (Morgan, 1986). But to the extent that many real-life networks display this property, and it is clearly a match to certain types of highly relevant problem, it remains highly worthwhile exploring the concept further.

Terrorist Organisations

A real-world example of contemporary organisations frequently analysed using network based approaches are those associated with terrorist activities. It is organisations such as these that in part are a cause and a consequence of NEC itself. In the present analysis two such organisations have been analysed, the network associated with the events of 9/11 (Krebs, 2001) and a similar network for Indian terror organisations (Basu, 2004). Note that in both cases the data was either constructed from publically available sources (in the case of the 9/11 networks from aggregated data based on news reports) or else made publicly available (in the case of Indian terror organisations via a published paper).

    These two terror organisations exhibit a power-law relationship consistent with the small-world paradigm, and when analysed for the characteristic pattern of high clustering (using betweeness centrality) and short path length (based on diameter) they do indeed meet this criteria. This is in comparison to both classic C2 or the theoretical maximum represented by the edge organisation, both of which do not. The networks and corresponding power laws are shown below in Figure 7.11 where this difference is clearly evident.

    In addition to the caveats above concerning small-world networks, a further caveat is attached to these real-world terror networks. Firstly, it is often difficult to discover the full extent of the network and it is likely that the analyst is only looking in on a part of it (although if it follows the small-world rubric then its scale-free nature means that it can be increased in size whilst retaining its distinctive properties). Also, the entry point into the network, into which associations and other linkages are constructed, may represent a part of the network materially different from the rest. Of course, in addition to all this, intelligence data may simply not capture sufficient information on interconnections, meaning that hubs may be represented in the network as outliers. Be that as it may, the two examples still enable some worthwhile comparisons to be made.

NEC Benchmarked Against Next Generation Organisations

Live-NEC, the small-world archetype, the decentralised with clusters network (a representation of the internet), and the two terror organisations were plotted

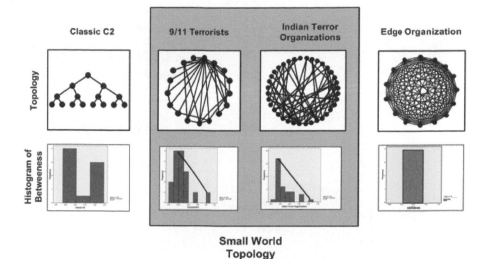

Figure 7.11 **Classic C2, Edge and terrorist organisations with their corresponding histograms showing the characteristic spread of nodes having high and low levels of connectivity. The terror networks exhibit 'small-world' properties**

into the NATO SAS-050 approach space as shown in Figure 7.12. What is clearly evident is that these various small-world networks cluster into a similar region to live-NEC and civilian C2, the region for which experimental and organisational data seemed initially lacking. As noted earlier in the chapter, this region of the approach space does indeed seem to contain networks with quite unique properties. These properties can now be stated. They are the properties associated with small-world networks. To that end, it is possible to add to the Aston Study's existing taxonomy and define this octant of the approach space accordingly.

In terms of live-NEC's relative position within the approach space, it is clear that while it, too, exhibits a certain power function relationship in terms of its connectivity (as shown in Figure 7.12), other factors either move it, or constrain it, such that it is positioned in the nearby 'personnel bureaucracy', 'effects based' octant (as described earlier).

What is interesting to note is that the people at work within that particular organisation played a significant part in moving it into its present region. As described in Chapter 6, the system ended up as being characterised by a number of significant emergent behaviours which lend weight to the fact that the people in the system were evolving it, and the interaction they needed, away from the anticipated trajectory which points towards the edge organisation ideal, and towards this small-world region of the model. The reason for venturing this hypothesis is that systems which have *evolved* quite often exhibit small-world properties. This, then, is one plausible explanation for the disparity between this particular version of live-NEC's anticipated versus actual movement within the approach space. The

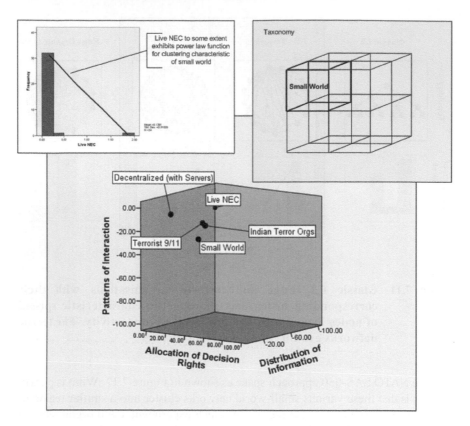

Figure 7.12   NATO SAS-050 approach space populated with net-enabled organisations and a live instance of NEC

users evolved it, and because it is evolved it exhibited something close to small-world properties. So, to the extent that small-world properties overlap with the aspirations of NEC (not to mention organisations with which it is tasked with dealing) it begs the question as to whether there are any explicit means of moving an organisation into this region, and what it means once it is there?

NEC versus Jointly Optimised Sociotechnical Organisations

Sociotechnical Organisations

Within the sociotechnical literature exists the now famous account of organisational redesign undertaken by Rice (1958) in textile mills in Ahmadabad, India. This provides a neat case study based example of an organisation explicitly redesigned according to sociotechnical principles.

Figure 7.13 shows that the redesign led to a radically different type of organisation within which, it was argued, the socio and the technical were jointly optimised. In other words, the real-life organisation benefited from the activity levels and cohesion seen in Leavitt's (1951) circle network, whilst possessing the stability and leadership of Leavitt's Star and Y networks. It possessed the hybrid nature of a personnel bureaucracy and the 'reorganisation was reflected in a significant and sustained improvement in mean percentage efficiency and a decrease in mean percentage damage [to goods]... the improvements were consistently maintained through-out a long period of follow up' (Trist, 1978, p. 53). Another effect that was noted was that: 'whereas the former organisation had been maintained in a steady state only by the constant and arduous efforts of management, the new one proved to be inherently stable and self correcting' (Trist, 1978, p. 53). It was also more satisfying to work within.

The defining feature of this redesign, something that is visually manifest in Figure 7.13, is the shift from a rationalistic style of hierarchical organisation to one based on smaller groups. There is a mixture of hierarchical subdivision and peer-to-peer interaction, of clusters and shortcuts. Hierarchical interaction is required so that task complexity, at the level of the entire system, can be managed. Peer-to-peer interaction is required to ensure a rapid response to local conditions without the need for more lengthy vertical interaction and effort on the part of higher management.

These so-called natural task groupings bestowed a form of autonomous responsibility resting on personnel within the system; there was now a 'whole task' and the requisite skills within the group to undertake it. The role of the group

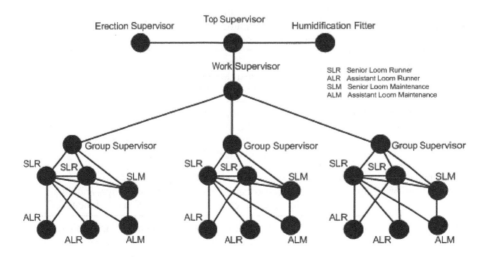

Figure 7.13   The organisation 'after' redesign. A simple organisation with complex tasks

Source: Trist, 1978.

leader was to work at the system boundaries (e.g., Teram, 1991), to 'perceive what is needed of him and to take appropriate measures' (p. 53). In command and control terms this new organisation shifts the primary task of managers away from processes of internal regulation to instead being focused on relating the total system to its environment (Trist, 1978). This is an important conceptual difference. Managers (or commanders) in this organisation became a form of executive, coordinating function, designing behaviours rather than scripting tasks, which of course is the aspect most often associated with being both 'arduous' and 'strenuous'.

This classic case study of organisational design appears to have many appealing analogues with contemporary visions of NEC, particularly in regard to network structures and the effects of them on the human actors they contain. We can also see rendered in this classic study some key attributes of sociotechnical system design made explicit in the work of Davis (1977). Table 7.1 presents a set of attributes inherent to jointly optimised sociotechnical systems.

Table 7.1   Attributes of jointly optimized sociotechnical systems

Systemic	'...all aspects of organisational functioning are interrelated'.
Open system	'...continuous adaptation to requirements flowing from environments'.
Joint optimization	The principle that socio and technical elements of an organisation should be jointly considered and maximised.
Organisational uniqueness	'...Structure of the organisation... suits the specific individual organisation's situation' (relates back to adaptation above).
Organisational philosophy	The design of structures and roles is 'congruent with agreed organisational values' (In other words, not a 'bolt-on' solution but pervasive and ubiquitous).
Quality of working life	'...integrity, values, and needs of individual members are reflected in the roles, structure, operations, and rewards of the organisation.' The intrinsic nature of work is enhanced (e.g., Hackman and Oldman, 1980).
Comprehensive roles for individuals or groups	The content of work and the people used to carry it out (and their organisation into teams or groups) should reflect the principles of 'meaningful' and 'whole tasks'.
Self-maintaining social systems	'...social systems are such that organisational units can carry on without external coercion... i.e., they are to become self-regulating'. This attribute relates well to Effects Based Operations as well as ad-hoc teams and flexible forces.

Table 7.1 *Concluded*

Flat structure	Although somewhat contrary to historical notions of military hierarchy, the attribute of a Sociotechnical System (one that is jointly optimised) is that there are 'fewer organisational layers or levels'.
Participation	'…democratization of the work place' with individuals able to contribute to problem solving and governance.
Minimal status differentials	This attribute seems to run counter to military thinking in terms of there being 'minimal differences in privileges and status', but on closer inspection it can be noted that any differences which are, 'unrelated to role and organisational needs' are regarded as divergent from a sociotechnical ideal.
Make large small	'Organisational and physical structures provide both a smaller, more intimate organisational boundaries and a feeling of smaller physical environment for individuals or groups'.
Organisational design process	'…components of the organisation evolve in a participative, iterative manner, only partially determined by advance planning'.
Minimal critical specification	This principle is (tacitly or otherwise) at the heart of Effects Based Operations. In organisational design terms,'… designers specify (design or select) the crucial relationships, functions, and controls, leaving to role-holders the evolutionary development of the remainder'.

Source: Davis, 1977, pp. 265–266.

NEC Benchmarked against a Sociotechnical Organisation

Jointly optimised organisations, those designed according to sociotechnical principles, achieve a multitude of complementary aims. In terms of the network archetype data, this sort of structure seems to create the conditions for cohesion and an improved experience for those at work within such systems. For commercial organisations, it seems to bestow some of the favourable aspects of small organisations on those that are large in size (see 'make large small' in Table 7.1 above). In a sense, then, the evolutionary cycle running from implicitly structured organisations, through bureaucracies of one sort or another, back to the recapturing of properties associated with these less formal organisational types is complete.

Despite sociotechnical systems theory being over 50 years old now, once again we find it just as applicable and relevant to NEC, whether referring to Leavitt's classic case studies, civilian examples of command and control, or even much more recent concepts such as small-world networks. Figure 7.14 shows that Rice's

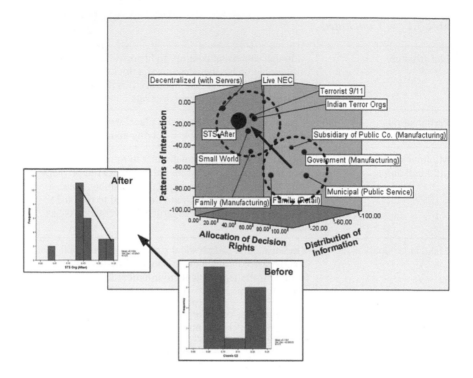

Figure 7.14   NATO SAS-050 approach space populated with pre- and post-sociotechnical organisation. Net-enabled organisations and the live instance of NEC are also shown (as faded points) to set sociotechnical systems in a modern context

(1958) application of sociotechnical principles to a real-life industrial concern causes its position to shift within the approach space from the more traditional region of manufacturing organisations towards the region occupied by small-world networks.

Sociotechnical systems theory, therefore, is one answer to the question of 'how can we bestow organisations with the desirable properties associated with NEC's opponents, collaborators in the civilian domain, and interesting theoretical networks such as small-worlds?'

Summary

Although representing a paradigm shift in the military arena, what could be regarded as NEC is alive and well in other domains. Populating the enhanced approach space with different military and non-military organisations enables the incipient state of NEC to be placed into some kind of context. In summary, when we speak of going Beyond NEC what the analysis presented in this chapter is driving at is the following:

- Although often mistaken for a theoretical ideal, it is clear that the Edge Organisation is a concept, as Alberts and Hayes (2005) originally note. What is clear is that the network structure that actually evolves is quite different to this;
- Many modern net-enabled network structures seem to gravitate towards a reasonably distinct part of the problem space. To the extent that this region matches the extant problem, the small-world characteristics associated with it appear useful and worth exploring.

One way of endowing organisations with these desirable small-world characteristics is to design them according to sociotechnical principles. This brings with it the further advantage of joint optimisation and an explicit set of HFI guidance and principles for achieving this end.

Chapter 8

The Design of Everyday Networked Interoperable Things

Aims of the Chapter

The preceding chapters have built to a progressively higher level of analysis, culminating in Chapter 7 in which the scale of observation is firmly rooted at the level of the total organisation. In this chapter we invert that scale and descend right down into the artefacts and entities that comprise the system's individual technical components, the scale of observation more normally associated with human factors. The contention is that as a consumer of such equipment the military (like all consumers) has probably become used to a dominant design paradigm; the closed, bureaucratic, inflexible, complex, technology-laden piece of kit which, despite all that, really only permits the user to perform simple and arbitrary individual tasks and often only then with arduous training and operational effort. This chapter attempts to shift that paradigm or, if not shift it, then at least put forward a case for an alternative. From the evolution of military equipment to its co-evolution with human users, from a focus on what equipment 'is' to what it actually 'does', an argument for the application of systems principles to the type of equipment now found in network enabled domains is developed. This enables a set of initial propositions posed at the beginning of the chapter to be backed up by theory and evidence and, in the next chapter (conclusions) to be elevated to the status of actionable design principles. Drawing widely from the domains of human factors and sociotechnical systems theory a case is thus put forward for equipment (and its procurement) to be as open, flexible, agile and self-synchronising as the net-enabled system into which it is designed to operate.

The Proposition(s)

This chapter begins by being deliberately contentious:

Proposition #1	Although it is technically convenient to see items of military equipment as stand-alone, none of it really exists in isolation.
Proposition #2	Equipment design and procurement is often based on a set of inappropriate implicit theories.

Proposition #3	It is not possible to achieve NEC's aspirations through top-down processes of design alone.
Proposition #4	It is not possible to specify all user requirements at the beginning of the design process. It is not even desirable.
Proposition #5	Design is not a one-off process and there is no clear end-product.
Proposition #6	Functionality is mostly split across artificial functional boundaries.
Proposition #7	The way that equipment should be used is often over-specified.
Proposition #8	Step changes in capability consistently fail to meet expectations.
Proposition #9	Design itself is often not designed.
Proposition #10	And the moment people start using equipment they are designing the next version of it.

These propositions could have been put forward in the delicately nuanced and heavily caveated manner which they ordinarily would deserve. We have chosen instead to frame them in deliberately contentious terms because the next task, which is to build a theoretical argument to support them, becomes considerably more challenging and meaningful. The only caveats that we do need to put in place are the following: firstly, these propositions relate to a particular class of networked, interoperable equipment that is becoming increasingly ubiquitous in twenty-first-century military environments: they by no means relate universally to all forms of equipment. Secondly, the perceptive reader will also note that these propositions and arguments are vaguely reminiscent of Cherns' (1976/1987), Davis' (1977) and Clegg's (2000) classic 'principles of sociotechnical design'. This is deliberate, too. Whilst we are certainly not the first to be as contentious as this, we do believe that this supporting work is as relevant now as it always has been. With the advent of NEC, probably more so.

In the sections that follow an alternative vision of how to think about and design this specific class of military network-enabled equipment is presented. A collection of explicit concepts and theories, some of them with a substantial legacy of practical application, suggest that there are fundamentally different ways of approaching the problem. We use the British Army's digital tactical communications system, called Bowman, as an example of how and why the propositions made at the beginning are closer to valid, actionable design principles than their contentious nature might at first suggest.

The Information Age

A number of attributes qualify the assertion that the Bowman communications system represents an incipient information age system. As a product Bowman is, in some senses, less about what it 'is' (i.e., a collection of green boxes and cables) but what it is connected to and what it 'does' (Kelly, 1994). From the user's point of view, Bowman represents a form of mobile porthole into a military 'blogosphere' populated by other people, information and assets. It should enable personnel to extract value from this collection of interconnected artefacts, to harness the capability that this provides in order to do meaningful 'Effects-Based' activities easily, only one of which is talking to people over the radio.

If flexibility, innovation and learning are the hallmarks of information-age equipment, then for military audiences used to considerably greater degrees of determinism this brings with it the appearance of an alarming lack of control. The key issue with this kind of networked, interoperable equipment, especially when combined with greater degrees of peer-to-peer working and effects based operations under the auspices of NEC, is that it creates the conditions for people to 'discover' ways of usefully deploying it. The more flexibility and ease-of-use, the more 'discovery' potential there is. This means that many of the ways in which current and future functionality will connect to what people want to do with it, that is to say the behaviour of such equipment, remains as yet undiscovered by users. Whilst the generic capability to interact and exchange information has been provided, what users decide to do with that capability, how they link it to the effects they want to achieve, will be up to them: this is the essence of self-synchronisation.

From the moment users put information-age equipment to use, the perceptive designer will see that a form of participatory, democratic design process is already underway. Thus in some senses Bowman, and other net-enabled equipment, is not an 'end product' at all, at least not in the traditional sense. What has been designed is often something more akin to a set of initial conditions or 'capabilities', a system that 'becomes' rather than a system that is frozen in time. An ongoing process of user/product evolution and co-evolution will tell exactly what. The following sections develop these basic premises further.

Design Evolution

End Products versus Initial Conditions

Equipment is not just manufactured: it is also 'designed'. Because it is designed it is subject to a range of diffuse interconnected influences, from competitive and commercial pressures to technology developments and user requirements. Equipment emerges out of a wider dynamic background and context; in a sense it too evolves, a form of natural process within which the designer plays a key role.

Like any system, Bowman has its own evolutionary timeline (Figure 8.1), its own inherited traits, its own 'equipment DNA' and its own adapted state vis-à-vis its environment; at least conceptually. Natural evolution, as distinct from the artificial evolution of equipment, is a bottom-up process. There is no 'control' or 'design' as such and complexity emerges out of simplicity. Implicit in bottom-up

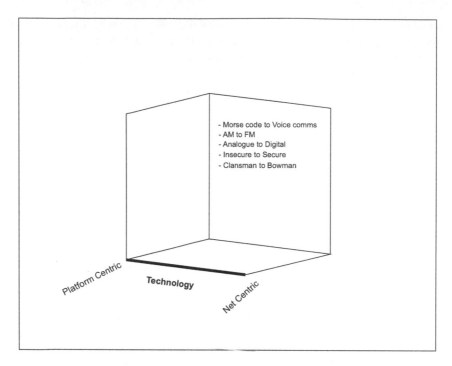

Figure 8.1   Bowman's (highly simplified) evolutionary timeline

processes like natural evolution is a subsumption architecture in which higher (complex) levels subsume lower (simple) levels. The rules of subsumption proposed by Brooks (1986) are instructive for equipment designers because they seem to map well onto the verb or capability-like properties of information age equipment such as Bowman:

- Step 1: Get the equipment doing simple things first and get them working perfectly.
- Step 2: Add new layers of activity over the results of the simple tasks.
- Step 3: Don't alter the simple things.
- Step 4: Make the new layer(s) work as perfectly as the layer below.
- Step 5: Repeat...

Bottom-up processes like these are not ideal in every circumstance. Neither are top-down processes. As Table 8.1 shows, design is contingent on the context of use and the type of problem a piece of equipment is meant to be solving.

The sort of artificial ecology that equipment normally resides in, and emerges from, exhibits a recognisable form of evolution but there is a combination of bottom-up and top-down processes. Achieving a balance between the two lies at the heart of Human Factors Integration (HFI) processes such as ISO 13407

Table 8.1   Matrix of 'Approach' versus 'Problem'

		Problem	
		Deterministic: i.e., stand-alone equipment	Complex: i.e., networked systems
Approach	Top-down	Matched	Rational systems start to behave irrationally...
	Bottom-up	Too slow and lacks scale...	Matched

(ISO, 1999). Achieving a balance is easier said than done and the more typical situation is a design ecology which is out of balance. Most common seems to be the application of long-standing top-down processes of design being used to solve complex non-deterministic problems. What typically arises from this mismatch are the technologically intensive pieces of equipment representative of the dominant paradigm (bureaucratic, inflexible, difficult to use, fails to meet aspirations etc.).

Opaque versus Transparent Capability

Let us now examine the bottom-up processes implicit in Bowman's evolutionary pre-history. This can be traced back as far as Morse code and the Crimean War. This was the first campaign to use electric telegraph and where it is sobering to learn that even then, the Commander-in-Chief received so many administrative queries from London that he was quickly overwhelmed with information. Morse code was gradually superseded by voice telephony (e.g., in the Boer War), voice telephony eventually led to radio-telephony (used to some extent in WWI) and onwards to the recognisably modern Larkspur radio system. This in turn was superseded by the lightweight, modern, but still ostensibly analogue Clansman system until the limitations of that created the conditions for voice and data communications under the aegis of Bowman.

Expressed in terms of subsumption there is an argument to suggest that Morse code first demonstrated the principle of electronic communications on any meaningful scale. This led to the nascent beginnings of a telecommunications-equipment infrastructure. This infrastructure, and the capability it afforded, created new uses and new aspirations for the system, which in turn paved the way for the next layer; voice telephony. This took the proven technology (of electrical signals carried by copper conductors) to the next level, enabling voice modulated signals to be carried rather than just dots and dashes. Again, this layer worked and served to create new affordances, affordances that helped the principle of voice telephony to break free from its wires through the use of radio, firstly AM (amplitude modulated and the bulky WWII era wireless sets) then VHF/FM and the post-war Larkspur and Clansman radios. In each case, communications spread further outwards from a 'strategic' mode of communication (e.g., the Crimean War and London communicating with field HQ) to 'tactical' (e.g., the Boer War, where field HQ used it to direct artillery fire). Presented in this way, Bowman's developmental pre-history is of course grossly overly simplified. It is intended merely as an illustration (not a detailed historical critique). The key point is that the continued outward spread of communications technology from strategic to tactical meant that the technology changed from being narrow, with specific users and highly defined uses, to pervasive, used by nearly every one for all manner of purposes. From platform-centric technology to net-centric.

This process continues. Some of the technology which is now a familiar part of military operations is itself becoming subsumed, a side effect of which is that it becomes increasingly transparent, 'weaving itself into the fabric of everyday life until indistinguishable from it' (Weiser, 1991, p. 94). Of course, the technology has not become 'literally' invisible, the point is that whilst it would be possible to point to and isolate the function that a specific cable or antenna serves, from the users perspective there is little point (Weiser, 1991). From the user's point of view the behaviour of the Bowman system has become largely disconnected from the specific technological artefacts that support it. It is the behaviour that counts. Despite its heterogeneous parts, the system as a whole not only works satisfactorily (as per Brook's subsumption rules above) but more importantly it behaves coherently (as per Actor Network Theory; Law, 2003). Only when the system breaks down does it dissolve into its constituent electronic components and human interventions, but even then this lack of coherency has more meaning for the signals engineer than it does for most Bowman users (Law, 2003).

Centralised versus Distributed Equipment

Technological invisibility goes hand in hand with another of Weiser's concepts: ubiquity. In practice what this means is that what a system like Bowman 'does' has not only been set free from the technology that supports it (i.e., the technology is transparent); equally important is that it has also been set free from the boundaries of space and location. Through systems like Bowman, information is becoming as 'dependable, consistent, and pervasive' as an electricity power grid (Chetty and Buyya, 2002, p. 61). As a result, information age equipment, whether it be something overtly 'radio-like' or something more complex like the various Bowman data terminals, all of it can be plugged in wherever this increasingly pervasive information infrastructure is present, from tanks to tents. Moving from left to right along Bowman's evolutionary axis, the difference in innovation and learning now potentially available to the user is akin to the kind of step change difference in the power of a product that runs off a battery compared to one that plugs into a mains supply.

Design Co-evolution

Stretched Capability

If technology is evolving, then so to are the users of it. The sociotechnical system, then, is 'co-evolving'. According to researchers in the field of Cognitive Systems Engineering (e.g., Hollnagel and Woods, 2005) technology and complexity are intertwined in precisely this way. In broad terms the extra utility afforded by some form of technological advance is usually seized thus 'pushing the system back to the edge of the performance envelope', rather like the motorway that is being continually widened and just as continually filled (Woods and Cook, 2002, p. 141). As a result, equipment tends to be run to its limits with all that that entails for reliability, stability and complexity (a bigger, wider motorway, at the level of the total system, is a more complex one; Hollnagel and Woods, 2005). The Law of Stretched Systems explains this self-reinforcing evolutionary cycle in relation to artefacts such as military equipment.

The cycle begins with an identified deficiency, a lack of capability, which is answered by expanding the equipment's functionality. Functionality is expanded by capitalising on the extra capability afforded by new technology, thus creating a new product which, like the wider motorway, is now a more complex one. Consider for a moment the functionality/ease of use afforded by the venerable Clansman radio (in which a curly cord to the handset was considered an innovation) and the functionality/ ease of use afforded by a Bowman data terminal? An attempt has been made to push the equipment 'back to the edge of the performance envelope' and to make the most of what technology now affords (Woods and Cook, 2002, p. 141). With extra capability has come greater task complexity which in turn creates new opportunities for problems and new deficiencies in capability, thus the cycle repeats.

An undesirable characteristic of this self-reinforcing cycle is that in capitalising on technology potential the user can often be left 'with an arbitrary collection of tasks and little thought may have been given to providing support for them' (Bainbridge, 1983, p. 151). In other words, the solution may be technically effective but not 'jointly optimised' with its human users. Because of this, human adaptability becomes required for equipment to work as intended which, in turn, creates new 'opportunities for malfunction'. Hollnagel and Woods clarify that 'by this we do not mean just more opportunities for humans to make mistakes but rather more cases where actions have unexpected and adverse consequences' (2005, p. 5). The typical response to this situation is to change the functionality of the system again. This completes the self-reinforcing cycle shown in Figure 8.2, which does not merely cause difficulties but represents an optimum strategy for maximising them (e.g., Norman, 1990).

Co-evolution

A well-known maxim in Human Factors is that 'it is easier to twist metal than it is to twist arms' (e.g., Sanders and McCormick, 1992). In other words, it is easier to

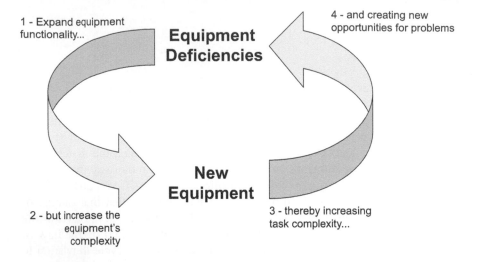

Figure 8.2   Hirshhorn's Law of Stretched 'Equipment'

Source: Hollnagel and Woods, 2005.

adapt equipment to its user than to rely on them adapting to it. When interpreted literally, it tends to presuppose that users do not readily change and that items of equipment can be seen in isolation from their environment.

Another way of looking at this twisting metal versus arms dialectic is to see it as an almost necessarily antagonistic process, such that there is 'reciprocal evolutionary change' or a little of both metal and arm twisting (Kelly, 1994, p. 74). Users have their arms twisted by having to adapt to new equipment, in turn, the equipment has a little more of its metal bent to suit new needs that arise from this adaptation, which creates more new needs, more arm twisting and more metal bending, on and on in a spiralling co-evolutionary fashion until the original piece of equipment becomes almost unrecognisable. As such, Bowman's evolutionary timeline says as much about what the technology has done to users as the users have done to the technology. Users and equipment have become locked more and more into a single system, 'each step of co-evolutionary advance winds the two antagonists more inseparably, until each other is wholly dependant on the other's antagonism. The two become one' (Kelly, 1994, p. 74; Licklider, 1960). Bowman's evolutionary pre-history provides an interesting example of this process in action.

The question to ask at the birth of recognisably modern military communications technology is, who actually 'needed' it? The historical answer is surprising. At the outset, and for many years following, relatively few people 'needed it' and it remained the more or less exclusive province of strategic communications at headquarters level. Its use as a tactical communications medium sprang from being able to direct flank formations and artillery fire, with this requirement in turn driven by the ability to undertake this sort of battlefield coordination without the use of wires. As radio technology improved, so did its reliability and resilience, and with it, the requirement for wired communications diminished. As technology improved still further, equipment became even more mobile, like Clansman. The interesting point is that none of these improvements fundamentally altered the nature of the communications task. It altered the context, the setting and the capability but the task of speaking into a receiver remained ostensibly the same. In equipment-design terms, Clansman was not a radical departure or paradigm shift from the antiquated field telephones and wireless sets that preceded it. It was not like Bowman, for example, that needs a critical mass of other Bowmanised bits of equipment in order to extract its full capability. It was merely a new layer of enhanced technology overlain on top of a proven method of working (i.e., radio communications).

As Clansman became more widespread and ubiquitous, as users and technology became increasingly locked into a single system, the metaphorical twisting of arms required more metal to be twisted. Enter Bowman. The step change in capability provided by Bowman derives from three areas of functionality: secure tactical communications, enhanced situational awareness (through global positioning technology) and a reliable data network. All this is designed to support the kind of interaction that users of Clansman (not to mention the internet, an interacting non-military trend) were coming to expect as they passed through the 'performance demands' phase of the self-reinforcing complexity cycle.

Reciprocal human/technological change continues. Within the heavily prescribed method of working embodied by the software suite (which resides on the various Bowman data terminals) a facility called Free Text is provided. This is nothing more or less than the ability for users to type text, then to send it across the data network to any other data terminal user (it is actually a secondary function embedded within a much larger super-ordinate capability). Because every communications eventuality seems to have been anticipated and subsequently incorporated into the software, giving rise to a highly specified method of working and an almost extreme level of functionality, it seems unlikely that the simplistic Free Text facility would be used all that often. After all, every template and pro-forma was provided so no-one really 'needed to'?

The technological metal of Clansman was bent into Bowman in response to new aspirations; users in turn are adapting Bowman technology in surprising ways. During a large-scale field trial (see Stanton et al., 2009) the effect of the simple Free Text facility became magnified out of all proportion. In a situation reminiscent of the explosion in SMS text messaging, it was observed that out of all the data communications events, 73 per cent of those initiated by the user were Free Text. This is surprising for a function that no one anticipated being used very often, if at all.

Step Changes

To paraphrase the classic sociotechnical systems literature: 'The single most descriptive term for [military] environments is change. This characteristic in itself is the basis for innovation of alternative [equipment], since the implicit assumption of [industrial age equipment] was high stability or placidity of the environment' (Davis, 1977, p. 263). The Larkspur radio handset, at one end of the evolutionary spectrum, has a simple, well-defined capability designed for an enduring context of use. It is an end product. Bowman, on the other hand, has the potential for through-life capability. Whether its innate flexibility and adaptiveness is seen explicitly as this or not, Bowman is designed for an altogether more dynamic environment, as 'a system designed to keep pace with technology' (MoD, 2008)

The problem with complex entities and environments is that they begin not to '…function in the linear ways in which we are used to thinking and analyzing.' (Smith, 2006, p. 40). 'actions are both persistent and strong enough to induce autochthonous processes in the environment' (Emery and Trist, 1965, p. 29). The self-reinforcing co-evolutionary cycle is one such autochthonous process, a type of positive feedback loop which means that 'the consequences which flow from […] actions lead off in ways that become increasingly unpredictable: they do not necessarily fall off with distance, but may at any point be amplified beyond all expectation [like Free Text]; similarly, lines of action that are strongly pursued may find themselves attenuated by emergent field forces [like overall system performance]' (Emery and Trist, 1965, p. 29). As a result, step changes in capability of the sort represented by Bowman are embarked upon with extreme caution. If the resultant system is not jointly optimised with users then the inevitable adaptations that they will perform will normally lead to the latter outcome rather than the former.

Interaction Pull and Technology Push

The graphical depiction of co-evolution and the shift to information age equipment shown in Figure 8.3, which is adapted from the work of Alberts, Gartska and Stein, (1999), represents a good summary for this section. Here it can be seen that an interactional y-axis has been added to Bowman's evolutionary timeline and the effect of co-evolutionary arm and metal twisting, of interaction and technology push, spirals forward in time. The interesting fact about this co-evolutionary spiral is that whilst it does indeed lead to complexity it does not necessarily lead to chaos: 'By incrementally extending new structure beyond the bounds of its initial state, [an information age system] can build its own scaffolding to build further structure […] with no bounds in sight' (Kelly, 1994, pp. 22–23).

Open Systems Behaviour

Object or System?

Formal systems thinking is '…a framework for conceptualizing or viewing the world' (Carvajal, 1983, p. 230). Although not always seen in this way, the special

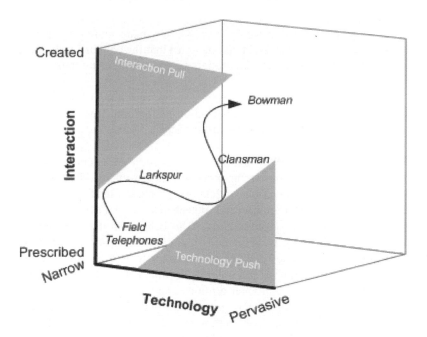

Figure 8.3   Interaction pull, technology push and equipment co-evolution
Source: Alberts, Garstka and Stein, 1999 p. 28.

case of networked, interoperable equipment can be seen as '…a set of interrelated elements' (Hall and Fagen, 1956 cited in Carvajal, 1983) and a 'regularly interacting or interdependent group of items forming a unified whole' (Merriam-Webster, 2003). Metcalfe's Law brings home the point behind looking at information age equipment in this way: 'as the number of [parts in a system] increases linearly the potential 'value or effectiveness' of the [system] increases exponentially' (Alberts et al., 1999). Information age systems like Bowman have more parts (not just Clansman-esque handsets but data terminals and more), more interconnections and potentially more value. The point of applying systems thinking to the type of equipment that lives in this networked environment is to try and harness such potential.

Objects versus Networks

The term 'network' has a very different meaning in systems theory than it might do in the world of NEC (where it is often attributed to the networked technology underlying it). In systems theoretic terms the extent to which a piece of equipment's 'parts' and 'interconnections' can be specified determines whether it has the systemic properties of an 'object' or a 'network'. The characteristics of an object bring to mind a relatively simple device such as the legacy Clansman radio handset. The characteristics of a 'network' are better aligned with the flexible, adaptable, information age attributes that technology like Bowman should offer. Looking at Bowman's evolutionary timeline it can be noted that the equipment on the left of the axis seems to exhibit object-like properties. They are, or tend to be:

- concerned with the attainment of a relatively specific goal,
- have well specified criteria for deciding on optimum means to ends, and
- have a 'high degree of formalization' (Scott, 1992).

According to Scott (1992) this is the definition of a closed or rational system. This is a system containing parts that have well-specified input/output characteristics and interconnections with known properties and flows. An electrical circuit diagram would be a good visual metaphor for such a system. The outputs of one component form the input to another, the behaviour of the component and the connection itself being well defined. It is often not just the technical parts of a product that are made to yield to closed systems thinking. The original Clansman handset, for example, has other well-defined input characteristics. Users lift the handset from its cradle, enter a number on the keypad and speak into the mouthpiece when they hear someone on the other end. The output characteristics are also definable, in so far as they are represented by the sound of a voice coming out of the earpiece. The first user is linked to the second user, functionally, by a simple two-way informational link. Obviously, it is possible to delve into greater detail but this is the essential essence of a Hierarchical Task Analysis (HTA) for this piece of equipment. A domestic telephone of similar design has around ten goals/operations in its HTA, which means for all practical purposes the use of a Clansman radio exploits the full, albeit limited, capabilities of the equipment and there is only one way to achieve an end state (which is the way the designer has provided). Equipment like this seems to make certain tacit assumptions about human users. The logic runs as follows:

- Rationality – the user, like the equipment, can be assumed to behave rationally. There is a well-defined end state and optimum prescribed ways of reaching those end states, which the user will follow logically and consistently.
- Linearity – 'the whole will be equal to the sum of the parts; […] the outputs will be proportionate to the inputs; […] the results will be the same from one application to the next; […] there is a repeatable, predictable chain of causes and effects.' (Smith, 2006, p. 40). This applies equally to both the human 'socio' elements of a system and its technical parts.
- Stability – end states, routes to end states, the context of use, the needs and preferences of users and the human system interaction remain static and enduring.

By only offering limited and simple functionality these assumptions are to some extent made tenable. It is certainly appropriate for the mono-functional

Clansman handset but raises important questions for the 'multi-functional' Bowman system. The HTA for the software system residing on Bowman's data terminals, for example, comfortably exceeds 300 goals/operations, not including an increasingly elaborate and expanding array of complex workarounds. Not only are there considerably more tasks but there are also more 'plans' that cue their enactment which, according to Annett's second principle of human performance, requires considerably more skill on the part of human users (e.g., Annett et al., 1971). However, the more complex the equipment becomes, and if it still adheres to the logic of simple machines, then the more it will have to rely on a prescribed form of human interaction. In practice, of course, it often yields a complex form of human adaptability in order to make it work as the designer intended. The real-world consequences of this are that what start out as highly rational products quite often degenerate into irrationality. From an equipment-design perspective, instead of remaining efficient, equipment rapidly degenerates into inefficiency as a result of its bureaucratic top-down design and poor usability. Systems then become unpredictable as users grow unclear about what they are supposed to do and do not get the outcome they expect. 'All in all, what were designed to be highly Rational [systems] often end up growing quite irrational' (Ritzer, 1993, p. 22). The hallmark of this can be seen in many large-scale projects, all of which meet their contractually enshrined requirements yet still exhibit paradoxical 'anti-synergistic' behaviour (e.g., Morris and Hough, 1987).

Information-age systems are different, or at least they should be. Here, users can do many things with the same piece of equipment, reaching the same end states from different initial conditions and in different ways. Information-age equipment is not concerned merely with the attainment of specific goals but also as yet unspecified ones. Information-age products should link users more closely to the kind of real-life 'effects based' tasks they want to perform, which means that if human adaptability is required then it is because of co-evolutionary needs rather than an artificial prescribed form of adaptability and workarounds. Rather than a circuit diagram, with known properties, moving up the vertical/structural axis from micro-systems to systems of systems, a more appropriate visual metaphor might be a block, Venn or influence diagram, one in which the properties and links are no less extant but more loosely specified. This type of equipment exhibits the systemic property of a network rather than an object.

Open Systems, Steady States and Equifinality

The idea of a network brings along with it several useful concepts, the first of which is that of the 'open system'. 'A system is closed if no material enters or leaves it; it is open if there is import and export and, therefore, change of the components' (Bertalanffy, 1950, p. 23). 'The 'open' perspective implies that the social and technological dimensions of [equipment] must be designed not only in relation to each other, but also with reference to evolving environmental demands' (Mitchell and Nault, 2003, p. 2). Open systems have boundaries with other systems and there is some form of meaningful exchange between them. An exchange that is not constrained by machine-like assumptions imposed upon human users.

'A closed system must, according to the second law of thermodynamics, eventually attain a time-independent equilibrium state, with maximum entropy and minimum free energy' (Bertalanffy, 1950, p. 23). A Clansman radio can exhibit 'time-independent states' with 'maximum entropy', at least conceptually. These high-level systems concepts make such a device look as if it is developmentally frozen; it performs one simple task in one simple environment, it cannot be changed or updated, there are no 'firmware upgrades', no plug-ins and no add-ons. With a real-life change in the environment from analogue to high capacity digital communications, the Clansman system couldn't inherently adapt so the British Army had to withdraw them and undertake a step-change to Bowman.

An open system, on the other hand, 'may attain (certain conditions presupposed) a time-independent state where the system remains constant as a whole... though there is a constant flow of the component materials. This is called a steady state' (Bertalanffy, 1950, p. 23). Steady state behaviour is an attribute of information age equipment and systems: 'They grow by processes of internal elaboration. They manage to achieve a steady state while doing work. They achieve a quasi-stationary equilibrium in which the enterprise as a whole remains constant, with a continuous 'throughput', despite a considerable range of external changes.' (Trist, 1978, p. 45). The behaviour and capability inherent in the various data terminals and other Bowman equipment is, to a significant degree, dependent upon the live, dynamic, information infrastructure that they are connected to. If Bowman was suddenly turned off, and with it the constant import and export of information, then all the data terminals would become closed systems and to all intents and purposes frozen and of limited use. Their capability exists as a steady state, a form of 'stable instability' (Kelly, 1994, p. 78) for which the following represents a new implicit design theory:

- Irrationality – 'people using the new [system] interpret it, amend it, massage it and make such adjustments as they see fit and/or are able to undertake' (Clegg, 2000, p. 467). They will adapt themselves and the equipment to suit their needs and preferences, which creates behaviour that is divergent from the normative, rational behaviour anticipated by designers (Hollnagel, 2005).
- Non-linearity – Industrial-age closed systems are often designed from the top-down. In systems terms, parts and interconnections are well defined and they are thus designed to be 'homopathic'; that is, the 'whole' is designed to be equal to the sum of the 'parts'. Information-age products can exhibit heteropathic effects, which means that they can become more than the sum of their parts. Capability, therefore, can be emergent and not traceable to any one cause or individual. To use Johnson's (2005) definition, these emergent properties are 'unexpected behaviours that stem from the interaction between the components [people] ...and their environment' (p. 1).

- Equifinality – end states, routes to end states, the context of use and the needs and preferences of users are dynamic and changeable. 'There are different ways of achieving the same purpose' (Majchrzak, 1997) from different initial conditions and by different means.

What information-age design is confronted with is not an either/or situation. The challenges of network-enabled system design can be partly explained by *both* the deterministic, industrial-age techniques of old just as much as they can by the probabilistic, information-age techniques of today (and the future). The key is ensuring that human factors approaches match the extant nature of problems.

The enduring dialectic throughout this chapter has been 'from' something 'to' something else. From 'platform centric' to 'net centric', from 'simplicity' to 'complexity', from 'linearity' to 'non-linearity'. If each of these transitions is ascribed an intersecting axis then a three-dimensional space is created that describes in more detail where the Army's tactical communications has come from and where it is heading to (Figure 8.4). One set of implicit theories, dominant design paradigms and conceptual languages applies to where tactical comms has been. The purpose so far has been to establish a foothold into the new implicit theories,

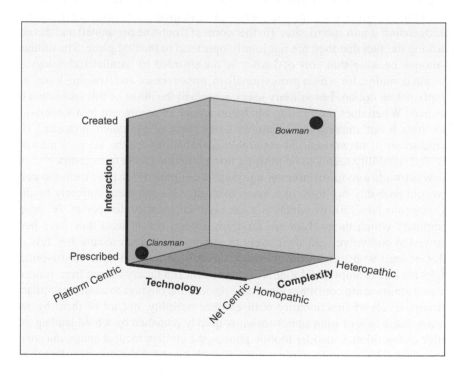

Figure 8.4   From Industrial to information-age products (based loosely on NATO, 2007)

emergent paradigms and conceptual languages applicable to where Bowman, and all information age equipment like it, is heading towards.

Summary

This chapter began with a deliberately contentious set of propositions which, given the arguments and evidence just presented, are perhaps rendered a little less contentious than they might have seemed at the beginning. The polemic tone of this chapter continues, however, with an observation that perhaps many readers will recognise. In the military sphere we have equipment often (but not always) built from scratch by specialist military suppliers, technically optimised through top-down processes of requirements capture (but again, not always), which themselves are rendered necessary because it is against these that the highly complex legal relationship between supplier and customer (i.e., the tax payer) is enshrined. The prevailing climate that has evolved is one in which there is nothing whatsoever in common with civilian equipment (i.e., products). Military equipment is serious.

Of course, problems still do arise and in this respect the military *is* unique. Here problems can be 'trained out'. This provides an artificially soft landing for many non-jointly optimised pieces of equipment, with military personnel 'bludgeoning it into submission' (in the words of front-line personnel) and thereby masking the fact that they are not jointly optimised in the first place. The military is unique because that sort of luxury is not afforded to parallel technology in civilian domains, for which professionalism, perseverance and 'training it out' are simply not an option. For military users, when will the limits of this adaptation be reached? When does the need to bludgeon a piece of equipment into submission and 'train it out' outstrip the capability a new piece of equipment provides? The introduction of networked, interoperable equipment under the aegis of network enabled capability seems to be pushing this adaptation closer to its limits.

What might a good information-age piece of equipment look like? Unfortunately, it would probably not look like many examples of equipment currently finding its way into NEC, all of which is a far cry from most contemporary PC based computers which themselves are far from perfect. But at least they have been allowed to co-evolve with their users to some extent, which means they rely on a lot of hard-won usage conventions (standardised left-and-right mouse-button clicks are one example). A lot of NEC equipment is a far cry indeed from some of the more pioneering consumer electronic devices that manage to combine similarly extreme levels of functionality with extreme usability, not all of them by any means, but a lack of joint optimisation is quickly punished by a hard landing and better competition. Consider mobile phones, the civilian tactical communications equivalent. The example might sound trivial but behind the brand attributes are serious pieces of technology with similar-sized HTAs to Bowman's data terminals. And all this contained in a highly portable device with a fraction of the button count and no opportunity (let alone desire) to train anything out. The new iPhone is a particularly trivial-sounding example, but here is an even more powerful device which dispenses with a keypad altogether and only has one button: on/off.

No one is suggesting that military equipment needs literally to look like this, or that there are very good reasons (e.g., security, durability, etc.) why military equipment is the way it is. But, as stated right at the start in Chapter 1, the paradigm is changing. There is a trend towards convergence, towards outward simplicity (built on subsumed and transparent inner complexity) and all of it rests on the idea that for all its vicissitudes the information age is not really, in itself, the problem. It is the design of the equipment that goes with it. The rate 'at which uncertainty overwhelms [a piece of equipment] is related more to its internal structure than to the amount of environmental uncertainty' (e.g., Carvajal, 1983). Sitter, Hertog and Dankbaar (1997) offer two strategies that can be applied to networked, interoperable military equipment.

'The first option is to restore the fit with the external complexity by an increasing internal complexity.' This is an acknowledged fact of Bowman. It is 'an extremely complex system that brings together a range of software functionality in a number of different hardware configurations. All of this in turn needs to be integrated with an array of platforms' (MoD, 2008). The alternative offered by a sociotechnical perspective is to: '...deal with the external complexity by *reducing* the internal control and coordination needs.' This option might be called the strategy of simple equipment that enables people to do complex, real-life, effects-based tasks. The paradox, then, is that a good information-age system is one that deals with external complexity *not* by a corresponding increase in *its* complexity (at least as far as the user is concerned) but by actually *reducing* complexity. All that has been discussed up until this point now comes to bear. The hallmark of information-age design is subsumption, transparent, ubiquitous and flexible technology – in a word, the application of open-systems principles to equipment which should be as 'self-synchronising' as the system within which it resides. To achieve this, design itself requires designing.

Chapter 9
Conclusions

'The power of NEC is derived from the effective linking or networking of knowledgeable entities [...].' (Alberts, Gartska and Stein, 1999, p. 6/7)

'These interactive technological and sociological patterns will be assumed to exist as forces having psychological effects in the life-space of [personnel]. Together, the forces and their effects constitute the psycho-social whole which is the object of study.' (Trist and Bamforth, 1951, p. 3/4)

'[...] some networks will be social, linking not computers and drones and Humvees but tribes, sects, political parties, even entire cultures. In the end, everything else is just data.' (Shachtman, 2007, p. 7)

Putting the Human Back into NEC

The 60 year legacy of sociotechnical systems theory reflects back a paradigm shift which the NEC community is starting to become aware of: 'You have your social networks and technological networks. You need to have both' (Shachtman, 2007). Sociotechnical systems theory is an answer to this call. It is an explicit strategy for putting the human back into complex, networked, interoperable, geographically dispersed, technologically mediated, large-scale systems like NEC. It sits at the junction of human-factors and complex-systems research, of engineering and design, of psychology, management and organisational science. Its strength (as well as that of this book) is not necessarily the depth of analysis it provides (one has to branch off into any one of the overlapping specialisms for that), but rather its interdisciplinary breadth. It does not speak exclusively of socio and/or technical parts, but of the interactions between them. This is something that the preceding chapters have focused on to deliver the following innovations:

- a mapping between the sixty year legacy of sociotechnical systems theory and the emergent world of NEC;
- a mapping between the fields of human factors and complex systems research;
- a mapping between classic organisational studies of the 1960s and the ability to benchmark twenty-first-century NEC organisations;
- a NATO SAS-050 model of command and control that can be populated with live data;

- a further domestication of systems concepts to make them applicable to the design of network centric military equipment.

On the one hand, whilst cross-disciplinary research of this sort is able to yield novel insights such as those listed above, on the other hand it is also fraught with difficulty. In drawing widely from domains not normally associated with human factors, a difficult balance has had to be struck between strict scientific rigour and innovation, breadth of coverage versus depth, concepts, theories, abstractions and analogies versus hard, inviolable facts. Written from a human-factors perspective it is unlikely that we have struck an optimum balance for all potential readers. Approaching this work from the classic sociotechnical school (of Human Relations and the Tavistock Institute) then doubtless some of the classic themes will be familiar, but the current work is of an undeniably different feel. The same is probably true (at least to some extent) for its more contemporary developments and extensions. As for military readers, we have probably provided something that is uncharacteristically unencumbered by in-depth military grounding (although we argue that this could be seen as a strength) and for complexity scientists the work is undoubtedly maths-light. Even human-factors readers will doubtless encounter much that is unfamiliar. This situation arises as an artefact of working at this difficult interface, but by declaring this openly we hope that suitable allowances can be made. The aim of this work is certainly not to antagonise or be deliberately contentious; rather, it is to offer the scientific community working in this area an extension to their domain-specific vocabularies, an insight into how human factors looks upon the challenges posed by NEC and a polemic that others can feel free to borrow from, draw parallels to, modify, criticise and argue with.

If military theorist B. H. Liddell Hart is reported as saying of the military that 'the only thing harder than getting a new idea into the military mind is getting the old one out' (Cebrowski and Gartska, 1998) then for the emergent world of NEC it is probably a case of 'getting the new idea *out there*'. Of course, it would be false pride on the part of human factors to assume that its ideas were the only ones. To be quite clear about this, we feel that human factors has as much to benefit from sociotechnical systems theory and the other fields that NEC brings it into contact with, as those other fields could benefit from human factors. The current volume, then, is little different from most other NEC texts. All of these come with a health warning, too.

NEC, not to mention many of its constituent fields of scientific endeavour, is not fully formed. It is dynamic, evolving, changeable, often incoherent, and to think of it in any other way is potentially misleading. Thus 'it will be decades before the real book on Network Centric Warfare will be written' noted Alberts, Gartska and Stein in 1999. A decade on and such a book is still waiting to be written – if indeed such a book can ever be written. We certainly cannot claim to have written anything approaching 'the real book on Network Centric Warfare' within these covers. What we have merely tried to do is respond to the first ten years of dynamism, evolution and change, and the evident need to start considering the social as well as the technical in NEC. This book is also 'designed to help prepare for the journey that will take us from an emerging concept to the fielding of real operational capability' (Alberts, Gartska and Stein, 1999, p. xiii) and we also 'do not pretend to have all of the answers, but we do feel that the ideas expressed here are worthy of your attention' (Alberts and Hayes, 2003, p. xxi).

Sociotechnical Principles for NEC System Design

The previous chapters provide a wide-ranging exploration of sociotechnical concepts and their relation to NEC. The following design principles flow from the previous chapter and enable the discussion to be summarised and pointed in the direction of practical application.

The principles owe a great deal to the work of Cherns (1976), Davis (1977) and Clegg (2000), all of whom have attempted such a distillation before, albeit in non-military domains. Design principles act as useful heuristic devices. They, too, are minimally critically specified. They are not about telling the designer what to do but about the mindset required to bring about solutions that convey the spirit of sociotechnical systems theory.

Three points are relevant. Firstly, the principles are as interconnected as the information-age design environment to which they are applied ('it would be bizarre if they were not'; Clegg, 2000, p. 464–465). They are, therefore, non-orthogonal. Secondly, they are prone to the same evolutionary and co-evolutionary forces as NEC itself; they, too, represent a set of initial conditions from which co-evolutionary scaffolding can continually be built. As a result, not only are they likely to express any current gaps in our understanding, they make no assumptions about the future either. This means that they are not likely to be complete. Thirdly, sociotechnical theory and NEC system design is not 'rendered non-problematic' (Clegg, 2000, p. 464) through their application. They, like the equipment and systems to which they are directed, are not an end product. In a very real sense the 'process' of working with the principles is probably of equal importance to the principle itself.

To the extent that they encapsulate something of the spirit of sociotechnical theory and NEC systems design, the following ten principles are offered (Table 9.1).

Visions of Success

It is important to state that the principles of sociotechnical design shown above do not have universal application. They may only appear so because the existing 'industrial age' paradigm is so difficult to adjust. Indeed, this dogged insistence on an implicit theory which is not appropriate in all situations has been a topic in the classical sociotechnical literature, organisational science and sociology for many

Table 9.1   Sociotechnical Principles of NEC System Design

Principle #1	NEC relies on open systems characteristics. Information becomes a commodity and there is constant import and export of it. The structure and type of these interactions is as much of a determinate of NEC's purpose and function as the physical system itself. This principle, therefore, is all about a shift in thinking from design being good at 'doing the parts' to design becoming good at 'doing the interconnections'. This in turn relies on **MULTI-DISCIPLINARY INPUT** and a recognition that **EQUIPMENT DOES NOT EXIST IN ISOLATION**.
Principle #2	There is a fundamental requirement to match design approaches/methods/ techniques to the fundamental nature of the problem/environment within which equipment will reside: **IMPLICIT THEORIES NEED TO BE TESTED**. Top-down approaches are appropriate for complicated, large-scale problems, bottom-up approaches are appropriate for complex, emergent problems. Sociotechnical systems theory and human factors integration is about achieving the correct balance.
Principle #3	'...design choices are contingent and do not necessarily have universal application' (Clegg, 2000, p. 468). What works in one situation and context may not work in another. Design choices may themselves have unintended consequences, creating effects that can become magnified or attenuated out of all proportion. In complex systems, one strategy for dealing with this is to use **BOTTOM-UP PROCESSES BASED ON SUBSUMPTION** (although see Principle #2).
Principle #4	The traditional conception of design is to respond to 'some articulated need' (e.g., Clegg, 2000, p. 466) yet, as we have argued, information age systems may embody 'needs' that will be subsequently discovered by users, users that may not even be the anticipated benefactors of the system. **USER REQUIREMENTS CO-EVOLVE** and will only unpack themselves over time.
Principle #5	Users of systems 'interpret it, amend it, massage it and make such adjustments as they see fit and/or are able to undertake' (Clegg, 2000, p. 467). NEC increases the opportunities for this adaptation as well as the speed with which this adaptation creates new co-evolutionary requirements. As such, **DESIGN FOR ADAPTABILITY AND CHANGE**.
Principle #6	A meaningful real-life task is one in which the user experiences a full and coherent cycle of activities, a task that has 'total significance' and 'dynamic closure' (Trist and Bamforth, 1951, p. 6). **DESIGN USEFUL, MEANINGFUL, EFFECTS-BASED WHOLE TASKS**.
Principle #7	'...one should not over-specify how a [system] will work [...] Whilst the ends should be agreed and specified, the means should not' (Clegg, 2000, p. 472). Here we are talking about an open, democratic, flexible type of technology that users can tailor to suit their own needs and preferences, in other words: **MINIMAL CRITICAL SPECIFICATION**.
Principle #8	Systems should be congruent with existing practices which may on occasion appear archaic compared to what technology now offers. Congruence capitalises on **HARD WON CO-EVOLUTION AND SYSTEM DNA**.
Principle #9	**USERS OR 'PROSUMERS'**: 'We, the users of the new system, are finding ways of exploiting its capabilities and thus helping you, the designers, to provide us with new capabilities'. From the moment users set NEC on the road to co-evolution, the perceptive designer will see that the design of future capabilities is already underway.
Principle #10	**DESIGN IS ITSELF AN INFORMATION-AGE ENTITY** and just as amenable to the same information age insights. There is clearly a paradox if NEC capability is being designed and procured according to 'industrial age' principles.

decades. Why do certain approaches continue to be applied when the evidence for their success is diminishing?

In this book we have argued that whilst there is not an overt conspiracy in favour of industrial-age approaches, the fact that this approach is rarely acknowledged to be as contingent as it really is, is a paradox that seems very hard to overcome. It is beyond the scope of this book to venture too far from the 'what' of this paradox into the 'why', but certainly one key reason seems to be the legal relationship that exists between supplier and customer. It is quite often within this relationship that certain end-products, deliverables, milestones and requirements are enshrined. Whilst processes such as Through-Life-Capability-Management and Smart Procurement are representative of attempts to shift this linear procurement paradigm the fact remains that for many nations NEC is over-budget, over-time and substantially less effective than originally intended (House of Commons, 2007). In other words, the approach *still* fails to match the problem. This particular problem/approach mismatch is pervasive and fundamental. Not least because human factors itself is sometimes just as guilty.

According to the varied literature in systems engineering and complexity science, the legacy of the Apollo lunar missions is partly responsible for this (e.g., Bar-Yam, 2004). Many of the project planning and systems engineering approaches which today are used to procure large-scale systems, like NEC, can quite often trace their lineage back to this era. The same can be said of human factors. Many of the methods and practices used today derive from large-scale systems such as nuclear power and process control, which were also culminating during what history increasingly refers to as the final phase of industry (e.g., Toffler 1980/1). In both cases, the legacy not only shapes the tools and techniques applied to current-day systems but also the aspirations people have for them.

Unfortunately, whilst it might be the case that the methodological zeitgeist of Apollo (or Polaris, nuclear power, process control, etc.) persists in the large scale projects that human factors practitioners work in to this day, one thing that has not persisted are their very dramatic successes. Stated simply, in spite of all the time, effort and expense that feeds into the design and development of what are now called 'megaprojects', the resultant organisations (e.g., Ritzer, 1993; Davis, 1977), systems (e.g., Bar-Yam, 2003) projects (e.g., Morris and Hough, 1987) and even NEC itself (House of Common, 2007), fail to meet the users', the designers', the procurers' and the public's aspirations. If this industrial-age zeitgeist needs rethinking then the good news is that so should NEC's aspirations. The proper context from which to judge its success is not against the legacy of technically optimised systems like Apollo but against those shown in Table 9.2.

As for the enduring image of Apollo and the set of expectations it evokes:

'In a scheme of history which has become the most popular plan of the recent past, the Space Age counts as the final phase of the Age of Industry – its culmination, just before the paradigm changed and the age of information replaced steel with digits.' (Spufford, 2003, p. 10)

Table 9.2   Large-scale project failures

System	Approx effort (years)	Approx. cost	Aspirations*	Outcome
Taurus computerised share trading system for the London Stock Exchange	3	£75m	'Paperless trading and computerized shareholdings [...] reduce time [...] bigger and better'	Scrapped
Channel Tunnel	6	£4.6bn	'One of the seven wonders of the modern world'	80 per cent cost overrun
Computerised dispatch system for the London Ambulance Service	1	£1.5m	'More efficient [...] automation [...] greater capacity'	Scrapped**
National Air Traffic Control system	<10	£339m	Existing system had 'functional limitations that would compel any modern engineer into laughter [...]'	Delayed and over-budget
NHS Computer System	10	£12bn	'A grandiose IT project [that would] transform the NHS'	Over-budget and reduced capability

* Aspirations derived from popular media via the BBC News website (www.bbc.co.uk/news).

** The failure was further implicated in the loss of 20 lives.

The Intervention of Last Resort

As Toffler's *Future Shock* (1981) and *Third Wave* (1980) predicted, the paradigm *has* changed. We are no longer in the realm of merely 'technical' problems that can be made to yield exclusively to linear, industrial-age methods of determinism, decomposition, hierarchy and reductionism. We are entering the zone of 'sociotechnical' problems, of non-linearity, co-evolution, emergence, change and dynamism. The promise of this new zone, into which the field of human factors itself seems to be pointing (see Chapter 4), is that the pursuit of joint optimisation brings with it the possibility of yielding disproportionately positive outcomes. As it is, when approaches and problems mismatch, when technical optimisation is achieved rather than joint optimisation, then seemingly rational systems start to behave irrationally. If NEC needs both the social and technical networks, if the object of study is the 'psycho-social whole', then it will be within the field of human factors that answers to these problems will be sought. This requires a set of tools, techniques and methods which match the extant nature of the problem, and a vocabulary that matches those of other allied disciplines. Human factors, often the intervention of last resort, will have to be ready to play a more proactive role.

References and Bibliography

Adler, N. and Docherty, P. (1998). 'Bringing business into sociotechnical theory and practice'. *Human Relations*, 51(3), 319–345.

Alberts, D. S. (1996). *The Unintended Consequences of Information Age Technologies*. NDU Press.

Alberts, D. S. (2003). 'Network centric warfare: Current status and way ahead'. *Journal of Defence Science,* 8(3), 117–119.

Alberts, D. S. (2007). 'Agility, focus, and convergence: The future of command and control'. *International C2 Journal*, 1(1), 1–30.

Alberts, D. S., Garstka, J. J. and Stein, F. P. (1999). *Network Centric Warfare: Developing and Leveraging Information Superiority*. CCRP.

Alberts, D. S. and Hayes, R. E. (2005). *Power to the Edge: Command and Control in the Information Age*. Washington DC: CCRP.

Alberts, D. S. and Hayes, R. E. (2006). *Understanding Command and Control*. CCRP.

Allen, R. E. (1984). *The Pocket Oxford Dictionary of Current English*. Oxford: Clarendon.

Amaral, L. A. N, Scala, A, Barthelemy, M. and Stanley, H. E. (2000). 'Classes of small world networks'. *Proceedings of the National Academy of Science, USA*. 97:11149–11152.

Annett, J. (2005). Hierarchical task analysis. In N. A. Stanton et al. (eds.), *Handbook of Human Factors and Ergonomics Methods*, (pp 33.1–33.7). London: CRC.

Annett, J., Duncan, K. D., Stammers, R. B. and Gray, M. J. (1971). *Task Analysis*. London: HMSO.

Apple (2007). Macworld San Francisco 2007 Keynote Address. Available at: http://www.apple.com/quicktime/qtv/mwsf07/

Arnold, J., Cooper, C. L., and Robertson, I. T. (1995). *Work Psychology: Understanding Human Behaviour in the Workplace*. London: Pitman.

Ashby, W. R. (1956). *Introduction to Cybernetics*. London: Chapman and Hall.

Atay, F. M. and Jost, J. (2004). 'On the emergence of complex systems on the basis of the coordination of complex behaviours of their elements'. *Complexity*, 10(1), 17–22.

Baber, C. and Houghton, R. J. (2008). Personal communication.

Baber, C., Houghton, R. J., McMaster, R., Salmon, P., Stanton, N. A., Stewart, R. J. and Walker, G. H. (2004). *Field Studies in the Emergency Services*. Aerosystems, Yeovil: HFI DTC.

Bainbridge, L. (1983). 'Ironies of automation'. *Automatica*, 19, 775–779.

Bainbridge, L. (1993). 'Types of hierarchy imply types of model'. *Ergonomics*, 36(11), 1399–1412.

Ball, P. (2004). *Critical Mass: How One Thing Leads to Another*. London: Arrow.

Barahona, M. and Pecora, L. (2002). 'Synchronization in small world systems'. *Phys. Rev.* Lett. 89, 054101.

Bartone, P. T. (2002). 'Factors influencing small-unit cohesion in Norwegian navy officer cadets'. *Military Psychology*, 14(1), 1–22.

Baruchi, I and Ben-Jacob, E. (2004). 'Functional holography of recorded neuronal networks activity'. *Neuroinformatics*, 2(3), 333–351.

Baruchi, I., Grossman, D., Volman, V., Shein, M., Hunter, J., Towle, V. and Ben-Jacob, E. (2006). 'Functional holography analysis: Simplifying the complexity of dynamical networks'. *Chaos*, 16, 015112.

Baruchi, I., Towle, V. L. and Ben-Jacob, E. (2005a). 'Functional holography of complex networks activity – from cultures to the human brain'. *Complexity*, 10(3), 38–51.

Baruchi, I., Towle, V. L. and Ben-Jacob, E. (2005b). 'Functional holography of bio-networks activity'. Presentation give to Gordon Conference on Nonlinear Science, Colby College, June 26–1 July.

Bar-Yam, Y. (1997). *Dynamics of Complex Systems*. Jackson, TN: Perseus.

Bar-Yam, Y. (2002). 'Complexity rising: From human beings to human civilization, a complexity profile', in *Encyclopaedia of Life Support Systems* (EOLSS). Oxford: EOLSS.

Bar-Yam, Y. (2003a). 'When systems engineering fails – toward complex systems engineering'. *International Conference on Systems, Man and Cybernetics*, (2)2021–2028, Piscataway, NJ: IEEE Press.

Bar-Yam, Y. (2003b). 'Complexity of military conflict: Multiscale complex systems analysis of littoral warfare'. Report: F30602–02–C-0158. Cambridge, MA: NECSI.

Bar-Yam, Y. (2004). *Making Things Work: Solving Complex Problems in a Complex World*. NECSI: Knowledge Press.

Bar-Yam, Y. (2004a). *Making Things Work: Solving Complex Problems in a Complex World*. Cambridge, MA: Knowledge Press.

Bar-Yam, Y (2004b). 'Multiscale complexity/entropy'. *Advances in Complex Systems*, 7, 47–63.

Bar-Yam, Y. (2004c). 'A mathematical theory of strong emergence using multiscale variety'. *Complexity*, 9(6), 15–24.

Basu, A. (2004). *Social Network Analysis of Terrorist Organisations in India*. Institute for Defence Studies and Analysis (IDSA).

Baxter, R. (2005). 'Ned Ludd encounters network-enabled capability'. *RUSI Journal*, Spring, 34–36.

Beekun, R. I. (1989). 'Assessing the effectiveness of sociotechnical interventions: Antidote or fad?' *Human Relations*, 42(10), 877–897.

Beringer, J. (1986). *The Control Revolution – The Technological and Economic Origins of the Information Society*. Cambridge: Harvard University Press.

Berman, M. (1983). *All That is Solid Melts into Air: The Experience of Modernity*. London: Verso.

Bertalanffy, L. V. (1950). 'The theory of open systems in physics and biology'. *Science*, 111, 23–29.

Beuscart, J. M. (2005). 'Napster users between community and clientele: The formation and regulation of a sociotechnical group'. *Sociologie du Travail*, 47, S1–S16.

Bevelas, A. (1948). 'A mathematical model for group structure'. *Applied Anthropology*, 7, 16–30.

Boehm, B. (2006). 'Some future trends and implications for systems and software engineering processes'. *Systems Engineering*, 9(1), 1–19.

Boguta, K. (2005). 'Complexity and the paradigm of Wolfram's A New Kind of Science'. *Complexity*, 10(4), 15–21.

Bowers, C. A., Jentsch, J., Salas, E. and Braun, C. C. (1998). 'Analyzing the communication sequences for team training needs assessment'. *Human Factors*, 40(4), 672–679.

Braard, P. O. (2001). 'Subjective task complexity and subjective workload: Criterion validity for complex team tasks'. *International Journal of Cognitive Ergonomics*, 5(3), 261–273.

Brand, S. (1974). *Cybernetic Frontiers*. New York: Random House.

Brooks, R. A. (1986). 'A robust layered control system for a mobile robot', *IEEE Journal of Robotics and Automation*, RA-2, April, 14–23.

Burke, M. A., Fournier, G. M. and Prasad, K. (2006). 'The emergence of local norms in networks'. *Complexity*, 11(5), 65–83.

Canas, J. J., Antoli, A., Fajardo, I. and Salmeron (2005). 'Cognitive inflexibility and the development and use of strategies for solving complex dynamic problems: effects of different types of training'. *Theoretical Issues in Ergonomics Science*, 6(1), 95–108.

Carvajal, R. (1983).' Systemic netfields: The systems' paradigm crises. Part I'. *Human Relations*, 36(3), 227–246.

Carvalho, P. V. R. d (2006). 'Ergonomic field studies in a nuclear power plant control room'. *Progress in Nuclear Energy*, 48, 51–69.

Carzo, R. and Yanouzas, J. N. (1969). 'Effects of flat and tall organisation structure'. *Administrative Science Quarterly*, 14(2), 178–191.

Cebrowski, A. K. and Gartska, J. H. (1998). 'Network-centric warfare – its origin and future'. *U.S. Naval Institute Proceedings Magazine*, 124(1/1,139), 1–9.

Chalmers, D. J. (1990). 'Thoughts on emergence'. Available at: http://consc.net/notes/emergence.html

Checkland, P. B. and Poulter, J. (2006). *Learning for Action: A Short Definitive Account of Soft Systems Methodology and its Use for Practitioners, Teachers and Students*. Chichester: Wiley.

Cherns, A. (1976). 'The principles of sociotechnical design'. *Human Relations*, 29(8), 783–792.

Cherns, A. (1987). 'Principles of sociotechnical design revisited'. *Human Relations*, 40(3), 153–162.

Chetty, M. and Buyya, R. (2002). 'Weaving computational grids: How analogous are they with electrical grids'. *Computing in Science and Engineering*, 61–71.

Chiles, W. D. (1958). 'Effects of elevated temperatures on performance of a complex mental task'. *Ergonomics*, 2(1), 89–96.

Ciborra, C., Migliarese, P. and Romano, P. (1984). 'A methodological inquiry of organisational noise in sociotechnical systems'. *Human Relations*, 37(8), 565–588.

Clark, D. M. (2005). 'Human redundancy in complex, hazardous systems: A theoretical framework'. *Safety Science*, 43, 655–677.

Clegg, C. W. (2000). 'Sociotechnical principles for system design'. *Applied Ergonomics*, 31, 463–477.

Coury, B. G. and Drury, C. G. (1986). 'The effects of pacing on complex decision-making inspection performance'. *Ergonomics*, 31(8), 1193–1203.

Cronshaw, S. F. and Alfieri, A. (2003). 'The impact of sociotechnical task demands on use of worker discretion and functional skill'. *Human Relations*, 56(9), 1107–1130.

Crutchfield, J. P. (1994). 'The calculi of emergence: Computation, dynamics and induction'. *Phys D*, 75, 11–54.

Cuevas, H. M., Costello, A. M., Bolstad, C. A. and Endsley, M. R. (2006). 'Facilitating distributed team collaboration.' Proceedings of the International Ergonomics Association (IEA) 16th World Congress on Ergonomics, Maastricht, The Netherlands, July 10–14, 2006.

Cummings, T. G., Molloy, E. S. and Glen, R. (1977). 'A methodological critique of fifty-eight selected work experiments'. *Human Relations*, 30(8), 675–708.

Czarniawska, B. and Hernes, T. (2005). *Actor-Network Theory and Organizing*. Copenhagen: Business School Press.

Dankbaar, B. (1993). *Economic Crisis and Institutional Change. The Crisis of Fordism from the Perspective of the Automobile Industry*. Maastricht: UPM.

Davis, L. E. (1977). 'Evolving alternative organisation designs: Their sociotechnical bases'. *Human Relations*, 30(3), 261–273.

Dawkins, R. (2006). *The Blind Watchmaker*. London. Penguin.

De Greene, K. B. (1980. 'Major conceptual problems in the systems management of human factors/ergonomics research'. *Ergonomics*, 23(1), 3–11.

Dekker, A. (2001). 'A category theoretic approach to social network analysis'. *Electronic Notes in Theoretical Computer Science*, 61, 1–13.

Dekker, A. (2002). 'Applying social network analysis concepts to military C4ISR architectures'. *Connections*, 24(3), 93–103.

Dodge, M. and Kitchin, R. (2001). *Atlas of Cyberspace*. London: Addison Wesley.

Driskell, J. E. and Mullen, B. (2005). 'Social Network Analysis'. In N. A. Stanton et al. (eds.), *Handbook of Human Factors and Ergonomics Methods* (58.1–58.6). London: CRC.

Elg, F. (2005). 'Leveraging intelligence for high performance in complex dynamic systems requires balanced goals'. *Theoretical Issues in Ergonomics Science*, 6(1), 63–72.

Emery, F. (1959). 'Characteristics of socio-technical systems,' reprinted in Emery, F. (1978), *The Emergence of a New Paradigm of Work*. Centre for Continuing Education, Australian National University, Canberra.

Emery, F. E. and Trist E. L. (1965). 'The causal texture of organisational environments'. *Human Relations*, 18 (1): 21–32.

Endsley, M. R. (1988). 'Situation awareness global assessment technique (SAGAT)'. Proceedings of the National Aerospace and Electronics Conference (NAECON). (New York: IEEE), 789–795.

Endsley, M. R. (1997). 'Situation awareness: The future of aviation systems'. Saab 60th Anniversary Symposium, Linkoping, Sweden, 8 September.

Ferbrache, D. (2005). 'Network enabled capability: Concepts and delivery'. *Journal of Defence Science*, 8(3), 104–107.

Festinger, L. (1950). 'Informal social communication'. *Psychological Review*, 57, 271–282.

Gladwell, M. (2000). *The Tipping Point: How Little Things Can Make a Big Difference*. London: Abacus.

Gleick, J. (1987). *Chaos: Making a New Science*. London: Sphere.

Grand, S. (2000). *Creation: Life and How to Make It*. Pheonix, London.

Granovetter, M. S. (1973). 'The strength of weak ties'. *American Journal of Sociology*, 78(6), 1360–1380.

Green, W. S. and Jordan, P. W. (1999). *Human Factors in Product Design: Current Practice and Future Trends*. London: Taylor and Francis.

Gregoriades, A. and Sutcliffe, A. G. (2006). 'Automated assistance for human factors analysis in complex systems'. *Ergonomics*, 49(12), 1265–1287.

Hackman, J. R. and Oldman, G. (1980). *Work Redesign*. New York: Addison-Wesley.

Hall, A. D. and Fagen, R. E. (1956). 'Definition of system'. General Systems, I.

Halley, J. D. and Winkler, D. A. (2008) 'Classification of emergence and its relation to self-organisation'. *Complexity*.

Hammarström, O. and Lansbury, R. (1991). 'The art of building a car: The Swedish experience re-examined'. *New Technology, Work and Employment*, 6(2), 85–90.

Hancock, P. A., Masalonis, A. J. and Parasuraman, R. (2000). 'On the theory of fuzzy signal detection: Theoretical and practical considerations'. *Theoretical Issues in Complexity Science*, 1(3), 207–230.

Hanisch, K. A., Kramer, A. F. and Hulin, C. L. (1991). 'Cognitive representations, control, and understanding of complex systems: A field study focusing on components of users' mental models and expert/novice differences'. *Ergonomics*, 34(8), 1129–1145.

Harary, F. (1994). *Graph Theory*. Reading, MA: Addison-Wesley.

Harris, C. J., and White, I. (1987). *Advances in Command, Control and Communication Systems*. London: Peregrinus.

Heller, F. (1997). 'Sociotechnology and the environment'. *Human Relations*, 50(5), 605–625.

Heylighen, F. and Joslyn C. (2001): 'Cybernetics and Second Order Cybernetics', in R. A. Meyers (ed.), *Encyclopaedia of Physical Science and Technology* (3rd edn), Vol. 4. New York: Academic Press, 155–170.

Hilborn, R. C. (2004). 'Sea gulls, butterflies, and grasshoppers: A brief history of the butterfly effect in nonlinear dynamics'. *American Journal of Physics*, 72(4), 425–427.

Hirschhorn, L., Noble, P. and Rankin, T. (2001). 'Sociotechnical systems in an age of mass customisation'. *Journal of Engineering and Technology Management*, 18, 241–252.

Hollnagel, E. (1993). *Human Reliability Analysis: Context and Control*. Academic Press: London.

Hollnagel, E. and Woods, D. D. (2005). *Joint Cognitive Systems: Foundations of Cognitive Systems Engineering*. London: Taylor and Francis.

Hornby, G. S. (2007). 'Modularity, reuse, and hierarchy: Measuring complexity by measuring structure and organisation'. *Complexity*, 13(2), 50–61.

Houghton, R. J., Baber, C., Cowton, M., Walker, G. H and Stanton, N. (2007). 'WESTT (Workload, Error, Situational Awareness, Time and Teamwork): An analytical prototyping system for command and control'. *Cognition Technology and Work*.

Houghton, R. J., Baber, C., McMaster, R., Stanton, N. A., Salmon, P., Stewart, R. and Walker, G. H. (2006). 'Command and control in emergency services operations: A social network analysis'. *Ergonomics*, 49(12–13), 1204–1225.

House of Commons (2007). 'Ministry of Defence: Delivering digital tactical communications through the Bowman CIP Programme'. House of Commons Committee of Public Accounts, Fourteenth Report of Session 2006–07, HC358. London: HMSO.

Howie, D. E. and Vicente, K. J. (1998). 'Measures of operator performance in complex, dynamic microworlds: Advancing the state of the art'. *Ergonomics*, 41(4), 485–500.

Hubler, A. W. (2005). 'Predicting complex systems with a holistic approach: The "throughput" criterion'. *Complexity*, 10(3), 11–16.

Hubler, A. W. (2006). 'Information engines: converting information into energy'. *Complexity*, 12(2), 10–12.

Hubler, A. W. (2007). 'Understanding complex systems'. *Complexity*, 12(5), 9–11.

ISO (1999). ISO 13407 Human centred design processes for interactive systems. Geneva: International Organization for Standardization.

Jenkins, D. P., Stanton, N. A., Salmon, P. M. and Walker, G. H. (2009). *Cognitive Work Analysis: Coping with Complexity*. Aldershot: Ashgate.

John, P. (2007). 'Contracting against requirements documents or shared models?' In Network Enabled Capability Through Innovative Systems Engineering (NECTISE) Conference, Loughborough, UK 12/13 Feb.

Johnson, C. W. (2005). *What are Emergent Properties and How Do They Affect the Engineering of Complex Systems?* University of Glasgow: Department of Computing Science.

Kaber, D. B., Riley, J. M., Tan, K-W and Endsley, M. R. (2001). 'On the design of adaptive automation for complex systems'. *International Journal of Cognitive Ergonomics*, 5(1), 37–57.

Kakimoto, T., Kamei, Y., Ohira, M. and Matsumoto, K. (2006). 'Social network analysis on communications for knowledge collaboration in OSS communities'. In *Proc. The 2nd International Workshop on Supporting Knowledge Collaboration in Software Development (KCSD'06)*, 35–41. Tokyo, Japan.

Kanter, R. M. (1983). *The Change Masters*. London: Unwin Hyman.

Karwowski, W. and Ayoub, M. M. (1984). 'Fuzzy modelling of stresses in manual lifting tasks'. *Ergonomics*, 27(6), 641–649.

Kazmar, M. M. (2006). 'Creation and loss of sociotechnical capital among information professionals educated online'. *Library and Information Science Research*, 28, 172–191.

Kelly, J. E. (1978). 'A reappraisal of sociotechnical systems theory'. *Human Relations*, 31(12), 1069–1099.

Kelly, K. (1994). *Out of Control: The New Biology of Machines, Social Systems, and the Economic World.* New York: Purseus.

Kirwan, B. (2000). 'Soft systems, hard lessons'. *Applied Ergonomics*, 31, 663–678.

Kleiner, B. M. (2006). 'Macroergonomics: Analysis and design of work systems'. *Applied Ergonomics*, 37, 81–89.

Knights, D. and McCabe, D. (2000). 'Bewitched, bothered and bewildered: The meaning and experience of teamworking for employees in an automobile company'. *Human Relations*, 53(11), 1481–1517.

Krebbs, V. E. (2001). 'Uncloaking terrorist networks'. *Connections*, 24(3), 1–12.

Law, J. (2003). 'Notes on the theory of the actor network: Ordering, strategy and heterogeneity'. Available at: http://comp.lancs.ac.uk/sociology/soc054jl.html

Leavitt, H. J. (1951). 'Some effects of certain communication patterns on group performance'. *Journal of Abnormal and Social Psychology*, 46, 38–50.

Lee, J. D. (2001). 'Emerging challenges in cognitive ergonomics: Managing swarms of self-organising agent-based automation'. *Theoretical Issues in Ergonomics Science*, 2(3), 238–250.

Lee, W., Karwowski, W., Marras, W. S. and Rodrick, D. (2003). 'A neuro-fuzzy model for estimating electromyographical activity of trunk muscles due to manual lifting'. *Ergonomics*, 46(1), 285–309.

Leplat, J. (1988). 'Task complexity in work situations'. In L. P. Goodstein, H. B. Andersen, and S. E. Olsen, (eds) *Tasks, Errors and Mental Models.* London: Taylor and Francis, 105–115.

Leweling, T. A. and Nissen, M. E. (2007). 'Hypothesis testing of edge organizations: laboratory experimentation using the ELICIT multiplayer intelligence game'. In *Proceedings International Command and Control Research and Technology Symposium*, Newport, RI.

Licklider, J. C. R. (1960). 'Man-computer symbiosis'. *IRE Transactions on Human Factors in Electronics*, HFE-1, 4–11.

Lo, S. and Helander, M. G. (2007). 'Use of axiomatic design principles for analyzing the complexity of human-machine systems'. *Theoretical Issues in Ergonomics Science*, 8(2), 147–169.

Luczak, H. and Ge, S. (1989). 'Fuzzy modelling of relations between physical weight and perceived heaviness: The effect of size-weight illusion in industrial lifting tasks'. *Ergonomics*, 32(7), 823–837.

Mael, F. A. and Alderks, C. E. (2002). 'Leadership team cohesion and subordinate work unit morale and performance'. *Military Psychology*, 5(3), 141–158.

Majchrzak, A. (1997). 'What to do when you can't have it all: Toward a theory of sociotechnical dependencies'. *Human Relations*, 50(5), 535–566.

Marmaras, N., Lioukas, S. and Laios, L. (1992). 'Identifying competences for the design of systems supporting complex decision-making tasks: A managerial planning application'. *Ergonomics*. 35(10), 1221–1241.

Martino, F. and Spoto, A. (2006). 'Social network analysis: A brief theoretical review of further perspectives in the study of information technology'. *PsychNology Journal*, 4(1), 53–86.

Masalonis, A. and Parasuraman, R. (2003). 'Fuzzy signal detection theory: Analysis of human and machine performance in air traffic control, and analytic considerations'. *Ergonomics*, 46(11), 1045–1074.

Mayo, E. (1949). *The Social Problems of an Industrial Civilization*. Routledge, London.

McCauley Bell, P. and Crumpton, L. (1997). 'A fuzzy linguistic model for the prediction of carpal tunnel syndrome risks in an occupational environment'. *Ergonomics*, 40(8), 790–799.

McLeod, R. W., Walker, G. H., and Moray, N. (2005). 'Analysing and modelling train driver performance'. *Applied Ergonomics*, 36, 671–680.

Merriam-Webster (2007). 'System'. Available at: http://www.merriam-webster.com/dictionary/system

Milgram, S. (1967). 'The small-world problem'. *Psychology Today* 1, 61–67.

Millhiser, W. P. and Solow, D. (2007). 'How large should a complex system be? An application in organisational teams'. *Complexity*, 12(4), 54–70.

Mintzberg, H. (1979). *The Structuring of Organizations*. Englewood Cliffs, NJ: Prentice-Hall.

Mitchell, V. L. and Nault, B. R. (2003). *The Emergence of Functional Knowledge in Sociotechnical Systems*. Haskayne School of Business, University of Calgary.

MoD (2005). *The BG PSO Combat Estimate*. Warminster: MoD.

MoD (2008). 'Bowman'. Available at: http://www.army.mod.uk/bowman

Moffat, J. and Manso, M. (2008). 'Development of the ELICIT model of informal networking to represent different NEC C2 maturity levels'. *TR*29294 v1.0, DSTL.

Molina, A. H. (1995). 'Sociotechnical constituencies as processes of alignment: The rise of a large-scale European information technology initiative'. *Technology in Society*, 17(4), 385–412.

Monk, A. and Howard, S. (1998). 'The rich picture: A tool for reasoning about work context'. *Interactions*, 5(2), 21–30.

Morgan, G. (1986). *Images of Organisation*. London: Sage.

Morris, P. W. G. and Hough, G. H. (1987). *The Anatomy of Major Projects*. Chichester: John Wiley and Sons.

Naikar, N., Moylan, A. and Pearce, B. (2006). 'Analysing activity in complex systems with cognitive work analysis: Concepts, guidelines and case study for control task analysis'. *Theoretical Issues in Ergonomics Science*, 7(4), 371–394.

NATO (2007). 'RTO-TR-SAS-050 Exploring New Command and Control Concepts and Capabilities'. Report: RTO-TR-SAS-050 AC/323(SAS-050)TP/50. NATO. Available at: http://www.rta.nato.int/Pubs/RDP.asp?RDP=RTO-TR-SAS-050

Niepce, W. and Molleman, E. (1998). 'Work design issues in lean production from a sociotechnical systems perspective: Neo-Taylorism or the nest step in sociotechnical design?' *Human Relations*, 51(3), 259–288.

Norman, D. A. (1990). 'The "problem" with automation: Inappropriate feedback and interaction, not "over-automation".' *Philosophical Transactions of the Royal Society of London*, B 327, 585–593.

Norman, D. A. (1998). *The Design of Everyday Things*. Cambridge, MA: MIT Press.

O'Brian, K. S. and O'Hare, D. (2007). 'Situational awareness ability and cognitive skills training in a complex real-world task'. *Ergonomics*, 50(7), 1064–1091.

O'Hare, D. (2000). 'The "wheel of misfortune": A taxonomic approach to human factors in accident investigation and analysis in aviation and other complex system'. *Ergonomics*, 43(12), 2001–2019.

Oliver, L. W., Harman, J., Hoover, E., Hayes, S. M. and Pandhi, N. A. (2000). 'A quantitative integration of the military cohesion literature'. *Military Psychology*, 11(1), 57–83.

Pasmore, W., Francis, C., Haldeman, J. and Shani, A. (1982). 'Sociotechnical systems: A north American reflection on empirical studies of the seventies'. *Human Relations*, 35(12), 1179–1204.

Patrick, J., James, N. and Ahmed, A. (2006). 'Human processes of control: Tracing the goals and strategies of control room teams'. *Ergonomics*, 49(12–13), 1395–1414.

Pava, C. (1986). 'Redesigning sociotechnical systems-design: Concepts and methods for the 1990s'. *Journal of Applied Behavioural Science*, 22 (3), 201–221.

Perrow, C. (1999). *Normal Accidents: Living with High-risk Technologies*. New Jersey: Princeton University Press.

Pondy, L. R. (1969). 'Effects of size, complexity, and ownership on administrative intensity'. *Administrative Science Quarterly*, 14(1), 47–60.

Pugh, D. S. and Hickson, D. J. (1976). *Organisation Structure in its Context*. Farnborough: Saxon House.

Pugh, D. S., Hickson, D. J. and Hinings, C. R. (1969). 'An empirical taxonomy of structures of work organisations'. *Administrative Science Quarterly*, 14(1), 115–126.

Pugh, D. S., Hickson, D. J., Hinings, C. R. and Turner, C. (1968). 'Dimensions of organisation structure'. *Administrative Science Quarterly*, 13(1), 65–105.

PWC, (2008). 'Today's Challenges'. Available at: http://www.pwc.com/extweb/f

Quesada, J., Kintsch, W. and Gomez, E. (2005). 'Complex problem-solving: A field in search of a definition?' *Theoretical Issues in Ergonomics Science*, 6(1), 5–33.

Reason, J. (1990). *Human Error*. Cambridge: Cambridge University Press.

Reber, A. S. (1995). *The Penguin Dictionary of Psychology*. London: Penguin.

Regan, G. (1991). *The Guinness Book of Military Blunders*. London: Guinness World Records:

Reinartz, S. J. (1993). 'An empirical study of team behaviour in a complex dynamic problem-solving context: A discussion of methodological and analytical aspects'. *Ergonomics*, 36(11), 1281–1290.

Resnick, P. (2002). 'Beyond bowling together: Sociotechnical capital'. In J. M. Carroll (ed.), *Human–Computer Interaction in the New Millennium* (647–672). Boston, MA: Addison-Wesley.

Reynolds, C. W. (1987). 'Flocks, herds, and schools: A distributed behavioural model'. *Computer Graphics*, 21(4), 25–34.

Rice, A. (1958). *Productivity and Social Organisation: The Ahmadabad Experiment*. London: Tavistock.

Richardson, M., Jones, G. and Torrance, M. (2004). 'Identifying the task variables that influence perceived object assembly complexity'. *Ergonomics*, 47(9), 945–964.

Rittel, H. and Webber, M. (1973). 'Dilemmas in a General Theory of Planning', *Policy Sciences*, 4, 155–169. [Reprinted in N. Cross (ed.), *Developments in Design Methodology*, J. Wiley and Sons, Chichester, 1984, 135–144.]

Ritzer, G. (1993). *The McDonaldization of Society*. London: Pine Forge Press.

Roetzheim, W. (2007). *Why Things are: How Complexity Theory Answers Life's Toughest Questions*. Jamul, CA: Level 4.

Rogalski, J. and Samurcay, R. (1993). 'Analysing communication in complex distributed decision-making'. *Ergonomics*, 36(11), 1273–1279.

Ropohl, G. (1999). 'Philosophy of socio-technical systems'. *Society for Philosophy and Technology*, 4(3), 1–10.

Roth, E., Scott, R., Deutsch, S., Kuper, S., Schmidt, V., Stilson, M. and Wampler, J. (2006). 'Evolvable work-centred support systems for command and control: Creating systems users can adapt to meet changing demands'. *Ergonomics*, 49(7), 688–705.

Rothrock, L., Koubek, R., Fuchs, F., Haas, M. and Salvendy, G. (2002). 'Review and reappraisal of adaptive interfaces: Toward biologically inspired paradigms'. *Theoretical Issues in Ergonomics Science*, 3(1), 47–84.

Ruddy, M. (2007). *ELICIT – The Experimental Laboratory for Investigating Collaboration, Information-sharing and Trust.* Parity Communications, Inc.

Salas, E., Bowers, C. A. and Cannon-Bowers, J. A. (1995). 'Military team research: 10 years of progress'. *Military Psychology*, 7(2), 55–75.

Salmon, P. M., Stanton, N. A., Walker, G. H. and Jenkins, D. P. (2009). *Distributed Situational Awareness*. Aldershot: Ashgate.

Sandberg, A. (1995). *Enriching Production: Perspectives on Volvo's Uddevalla Plant as an Alternative to Lean Production.* Aldershot: Avebury.

Sanders, M. S. and McCormick, E. J. (1992). *Human Factors in Engineering and Design*. Maidenhead, UK: McGraw-Hill.

Sarter, N. B. and Woods, D. P. (1997). 'Team play with a powerful and independent agent: Operational experiences and automation surprises on the airbus A-320'. *Human Factors*, 39, (4), 553–569.

Sauer, J., Felsig, T., Franke, H. and Ruttinger, B. (2006). 'Cognitive diversity and team performance in a complex multiple task environment'. *Ergonomics*, 49(10), 934–954.

Sauer, J., Hockey, G. R. J. and Wastell, D. G. (2000). 'Effects of training on short and long-term skill retention in a complex multiple-task environment'. *Ergonomics*, 43(1), 2043–2064.

Scacchi, W. (2004). 'Socio-technical design' in W. S. Bainbridge (ed.), *The Encyclopaedia of Human Computer Interaction.* Location: Berkshire Group.

Scott, R. (1992). *Organisations; Rational, Natural, and Open Systems*. New Jersey: Prentice Hall.

Seddon, J. (2003). *Freedom from Command and Control: A Better Way to Make the Work Work*. New York: Vanguard.

Shachtman, N. (2007). 'How technology almost lost the war: In Iraq, the critical networks are social – not electronic'. *Wired Magazine*, 15.12

Shannon, C. E. (Jul and Oct 1948), 'A mathematical theory of communication'. *Bell System Technical Journal*, 27, 379–423 and 623–656.

Shorrock, S. T. and Straeter, O. (2006). 'A framework for managing system disturbances and insights from air traffic management'. *Ergonomics*, 49 (12–13), 1326–1344.

Siebold, G. L. (2000). 'The evolution of the measurement of cohesion'. *Military Psychology*, 11(1), 5–26.

Siebold, G. L., and Kelly, D. R. (1988). 'Development of the combat platoon cohesion questionnaire (Tech. Rep. No. 817)', Alexandria, VA: U.S. Army Research Institute for the Behavioural and Social Sciences.

Sitter, L. U., Hertog, J. F. and Dankbaar, B. (1997). 'From complex organisations with simple jobs to simple organisations with complex jobs'. *Human Relations*, 50(5), 497–536.

Smith, E. A. (2006). *Complexity, Networking, and Effects-based Approaches to Operations*. CCRP Publication Series.

Solomonoff, F. (1960). *A Preliminary Report on a General Theory of Inductive Inference*, Report V–131, Cambridge, MA: Zator Co.

Spufford, F. (2003). *Backroom Boys: The Secret History of the British Boffin*. London: Faber and Faber.

Stanton, N. A. (2006). 'Hierarchical task analysis: Development, applications and extensions'. *Applied Ergonomics*, 37, 55–79.

Stanton, N. A., Baber, C. and Harris, D. (2007). *Modelling Command and Control: Event Analysis of Systemic Teamwork*. Aldershot: Ashgate.

Stanton, N. A., Jenkins, D. P., Salmon, P. M., Walker, G. H., Revell, K. and Rafferty, L. (2009). *Digitising Command and Control: Human Factors and Ergonomics Analysis of Mission Planning and Battlespace Management*. Aldershot: Ashgate.

Stanton, N. A., Salmon, P. M., Walker, G. H., Baber, C. B. and Jenkins, D. P. (2005). *Human Factors Methods: A Practical Guide for Engineering and Design*. London: Ashgate.

Storr, J. (2005). 'A critique of effects-based thinking'. *RUSI Journal*.

Suh, N. P. (2007). 'Ergonomics, axiomatic design and complexity theory'. *Theoretical Issues in Ergonomics Science*, 8(2), 101–121.

Swain, A. D. (1982). 'Modelling of human performance in complex systems with emphasis on nuclear power plant operations and probabilistic risk assessment'. *Ergonomics*, 25(6), 449.

Tapscott, D. and Williams, A. D. (2006). *Wikinomics: how mass collaboration changes everything*. New York: Portfolio

Teram, E. (1991). 'Interdisciplinary teams and the control of clients: A sociotechnical perspective'. *Human Relations*, 44(4), 343–357.

Thompson, P. and Wallace, T. (1996). 'Redesigning production through teamworking: Case studies from the Volvo Truck Corporation'. *International Journal of Operations and Production Management*, 16(2), 103–18.

Toffler, A. (1980). *The Third Wave*. London: Pan.

Toffler, A. (1981). *Future Shock: The Third Wave*. New York: Bantam.

Trist, E. L. (1978). 'On socio-technical systems'. In Pasmore, W. A. and Sherwood, J. J. (eds), *Sociotechnical Systems: A Source*book. San Diego, CA: University Associates.

Trist, E. and Bamforth, K. (1951). 'Some social and psychological consequences of the Longwall method of coal getting'. *Human Relations*, 4, 3–38.

Turner, A. W. (2008). *Crisis? What Crisis? Britain in the 1970s*. London: Aurum.

Ullman, H. and Wade, J. Jr. (1996). *Shock and Awe: Achieving Rapid Dominance*. National Defense University.

Venda, V. F. (1995). 'Ergodynamics: Theory and applications'. *Ergonomics*, 38(8), 1600–1616.

Verrall, N. G. (2006). 'Multinational experiment 4: UK analysis of the operational-level headquarters organisation'. *Defence Science and Technology Laboratory/ TL*18781 V0.1, 24th March 2006.

Vicente, K. J. (1999). *Cognitive Work Analysis: Toward Safe, Productive, and Healthy Computer-based Work*. Mahwah, NJ: Lawrence Erlbaum Associates

Vicente, K. J., Mumaw, R. J. and Roth, E. M. (2004). 'Operator monitoring in a complex dynamic work environment: A qualitative cognitive model based on field observations'. *Theoretical Issues in Cognitive Science*, 5(5), 359–384.

Viegas, F. B., Wattenberg, M. and McKeon, M. (2007). 'The Hidden Order of Wikipedia'. In D. Schuler (ed.) *Online Communities and Social Computing*, HCII 2007, LNCS 4564, 445–454. Berlin: Springer-Verlag:

Waldrop, M. M. (1992). *Complexity: The Emerging Science at the Edge of Order and Chaos*. New York: Simon and Schuster.

Walker, G. H., Gibson, H., Stanton, N. A., Baber, C., Salmon, P., and Green, D. (2006). 'Event analysis of systemic teamwork (EAST): A novel integration of ergonomics methods to analyse C4i activity'. *Ergonomics*, 49(12–13), 1345–1369.

Walker, G. H., Stanton, N. A., Salmon, P. M. and Jenkins, D. P. (2008b). 'A review of sociotechnical systems theory: A classic Concept for New Command and Control Paradigms'. *Theoretical Issues in Ergonomics Science*, 9(6), 479–499.

Walker, G. H., Stanton, N. A., Salmon, P., Jenkins, D. P., Monnan, S. and Handy, S. (2008c). 'An evolutionary approach to network enabled capability'. *International Journal of Industrial Ergonomics*, in press.

Walker, G. H., Stanton, N. A., Salmon, P. M., Jenkins, D., Revell, K. and Rafferty, L. (2008a). 'Extending social network analysis: Static and dynamic modelling of a live NCW case study'. *International C2 Journal*, in press.

Walker, G. H., Stanton, N. A., Stewart, R., Jenkins, D., Wells, L., Young, M. S., Salmon, P. and Baber, C. (in press). 'Using an integrated methods approach to analyse the emergent properties of military command and control'. *Applied Ergonomics*.

Walker, G. H., Stanton, N. A., Wells, L., Gibson, H., Young, M. S and Jenkins, D. (2008d). 'Is a picture (or network) worth a thousand words: Analysing and representing distributed cognition in air traffic control systems'. *Theoretical Issues in Ergonomic Science*, in Press.

Warm, J. S., Dember, W. N., and Hancock, P. A. (1996). 'Vigilance and workload in automated systems'. In R. Parasuraman and M. Mouloua (eds), *Automation and Human Performance: Theory and Applications*. (183–200). Mahwah, NJ: Lawrence Erlbaum Associates.

Waterson, P. E., Older Gray, M. T., and Clegg, C. (2002). 'A sociotechnical method for designing work systems'. *Human Factors*, 44(3), 376–391.

Watts, D. J. and Strogatz, S. H. (1998). 'Collective dynamics of 'small-world' networks'. *Nature*, 393(4), 440–442.

Weber, M. (1930). 'The protestant ethic and the spirit of capitalism'. e-book available at: http://www.ne.jp/asahi/moriyuki/abukuma/weber/world/ethic/pro_eth_frame.html

Wei, X. (2007). 'Identifying problems and generating recommendations for enhancing complex systems: Applying the abstraction hierarchy framework as an analytical tool'. *Human Factors*, 49(6), 975–994.

Weiser, M. (1991). 'The computer for the 21st century'. *Scientific American*, Sept, 94–100.

Weisstein, E. W. (accessed on 17th Sept 2008). 'Graph Diameter. From MathWorld-A Wolfram Web Resource'. Available at: http://mathworld.wolfram.com/GraphDiameter.html

Wikipedia (2007). 'Short message service'. Available at: http://en.wikipedia.org/wiki/Short_message_service

Wilson, J. R. (2000). 'Fundamentals of ergonomics in theory and practice'. *Applied Ergonomics*, 31, 557–567.

Woo, D. M. and Vicente, K. J. (2003). 'Sociotechnical systems, risk management, and public health: comparing the North Battleford and Walkerton outbreaks'. *Reliability Engineering and System Safety*, 80, 253–269.

Woods, D. and Dekker, S. (2000). 'Anticipating the effects of technological change: A new era of dynamics for human factors'. *Theoretical Issues in Ergonomics Science*, 1(3), 272–282.

Woods, D. D. and Cook, R. I. (2002). 'Nine steps to move forward from error'. *Cognition Technology and Work*, 4(2), 137–144.

Woods, D. D. (1991). 'The Cognitive Engineering of Problem Representations'. In G. R. S. Weir and J. L. Alty (eds), *Human-Computer Interaction and Complex Systems*, Academic Press, 169–188

Woods, D. D. and Dekker, S. W. A. (2000). 'Anticipating the effects of technological change: A new era of dynamics for human factors'. *Theoretical Issues in Ergonomic Science*, 1(3),272–282.

Woods, D. D. (1988). 'Coping with complexity: The psychology of human behaviour in complex systems'. In L. P. Goodstein, H. B. Andersen, and S. E. Olsen, (eds) *Tasks, Errors and Mental Models*. London: Taylor and Francis, 128–148.

Index